The Moral Lives of Animals

BY THE SAME AUTHOR

A Mad People's History of Madness

Big Things from Little Computers

Genesis II

Intelligent Schoolhouse

CoCo Logo

The Deluge and the Ark

Visions of Caliban (with Jane Goodall)

Chimpanzee Travels

Demonic Males (with Richard Wrangham)

Storyville, USA

Africa in My Blood (editor)

Beyond Innocence (editor)

Eating Apes

Jane Goodall

Elephant Reflections

The Moral Lives of Animals

DALE PETERSON

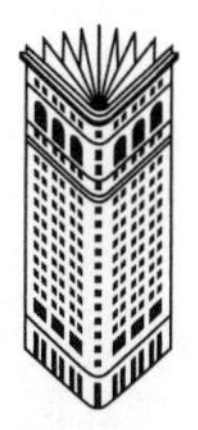

BLOOMSBURY PRESS
New York • Berlin • London • Sydney

Published by Bloomsbury Press, New York

All papers used by Bloomsbury Press are natural, recyclable products made from wood grown in well-managed forests. The manufacturing processes conform to the environmental regulations of the country of origin.

LIBRARY OF CONGRESS CATALOGING-IN-PUBLICATION DATA

Peterson, Dale.
The moral lives of animals / Dale Peterson.—1st U.S. ed.
p. cm.
Includes bibliographical references.
ISBN 978-1-59691-424-7 (hardcover)
1. Animal psychology. 2. Animal behavior. 3. Ethics.
4. Moral motivation. I. Title.
QL785.P375 2010
156—dc22
2010024662

First U.S. edition 2011

1 3 5 7 9 10 8 6 4 2

Typeset by Westchester Book Group
Printed in the United States of America by Quad/Graphics, Fairfield, Pennsylvania

This book is dedicated to:
Jarold Ramsey
Thomas C. Moser
Peter Matson

Contents

PART I

Where Does Morality Come From? Concepts

CHAPTER 1
Words

"Vengeance on a dumb brute!" cried Starbuck,
"that simply smote thee from blindest instinct!
Madness! To be enraged with a dumb thing,
Captain Ahab, seems blasphemous."

—Herman Melville, *Moby-Dick*[1]

Elephants can be dangerous. I remember thinking those very words with an unusual concentration one afternoon, while being chased by wild forest elephants in an impenetrable thicket in Ivory Coast, West Africa.

Well, the thicket was "impenetrable" from my perspective. From the elephants' perspective it must have been a minor nuisance, and, in truth, that contrast in perspectives was what made me so anxious.

I would like to say "I ran," but really I was just trying to push and drag my suddenly weak body through a stubbornly resistant barrier of vegetation.

"Chased"? It seemed so at the time. In retrospect, the elephants may not have been chasing me. I saw nothing. I heard the thudding of big feet hitting the earth and the crackling and crashing of large bodies hurtling through vegetation. But since I'm still alive and able to write these words, it may be that instead of chasing, they were, as I was, fleeing—having just sensed strange people creeping into their private thicket.

"Dangerous"? Here's how an elephant is likely to do it. He or she might knock you down, or maybe just toss you down with a grab and flick of the trunk, then stab you with a tusk, pin and crush you with a foot, or press down with that boulder-size forehead until you pop open like a piece of rotten fruit. Being inside a car is better, but an adult elephant, male or female, can run a tusk right through the door of your car or use a few tons of body weight to crush down from the top. It can't be a pleasant experience sitting inside that car, and in the end you'll consider yourself lucky merely to be alive and still able to articulate the words that tell what happened—assuming, of course, that you are.[2]

Still, no one is particularly surprised to hear that an elephant or any other wild animal is dangerous. Wild animals are supposed to be dangerous. It is surprising, though, when a wild animal deliberately seeks you out, seems to be pursuing you not out of some irrational explosion of rage, not from dumb and blind instinct, not according to an automatic, machinelike sequence of predatory behaviors, but rather with what looks like real intent and even, possibly, focused calculation.

Such may have characterized an encounter biologist Douglas Chadwick experienced one evening at the edge of the Nilgiri Reserve in southern India. In his book *The Fate of the Elephant* (1992), Chadwick describes the start of that evening in idyllic terms. After visiting the distinguished elephant expert Raman Sukumar, Chadwick began a pleasant late-afternoon hike with two young students who served as Sukumar's research assistants.[3]

Sukumar had warned Chadwick about the Nilgiri elephants, who were known to be particularly aggressive toward people. But the three hikers were passing along the edge of the reserve, a relatively open area where the trees thinned out and mixed with grass and shrub, and where, at the moment, many flowers were brilliantly blossoming in response to recent rains. Chadwick saw numerous chital (dappled Indian deer) and a couple of blackbucks. And as the cool of the coming dusk descended, the insects rose and, with them, the birds: cuckoos, hoopoes, magpies, mynahs, peacocks. Yes, it was all very lovely, but soon the fading light reminded the trio that they were still four miles away from their intended rendezvous, a spot on the road where a friend would be waiting with a car. They picked up the pace.

By the time they approached the dim strip of road and a dark hulk that seemed to be the waiting car, evening had arrived. Chadwick was carrying a flashlight, which he now flicked on as a friendly beacon to the driver. Immediately, however, a great burst of trumpeting shattered the peace. Chadwick turned off the light, heard and felt the thudding of heavy feet, and he and his two companions ran for their lives. They tried veering back in the direction of the road and car—only to be cut off by another burst of trumpeting and more pounding footfalls. They kept running. One of Chadwick's companions shouted at him to move in a zigzag pattern among the trees. (Because of their great mass, elephants have trouble making quick turns.) The American began zigzaggingly tripping among the denser shafts of darkness that must have represented trees, while still listening to, indeed *feeling*, that heavy thudding behind him. After a time, the pounding in the earth became indistinguishable from the pounding of his pulse. He stopped to listen and heard nothing. Of course, elephants, with their thickly padded foot-bottoms, can be extremely quiet when they want to be, and the biologist began to think he was being not chased so much as tracked. He began to feel, as he put it, "like elephant prey."[4]

He and his companions were afraid to return in the direction of the car, so finally they ran across to a different part of the road, flagged down a late-running bus, and in a small village tavern persuaded an inebriated car-owner to ferry them back to their waiting friend on the road. The friend was shaken, upset. The elephant had moved up beside the car in complete silence, so the driver inside, sitting next to his open window, had been about as startled and alarmed by the first explosive burst of trumpeting as Chadwick and his companions had been.

"I will never know what that elephant had in mind that night," Chadwick writes, "but upon reflection, I have to credit the animal with giving us fair warning. If it had really been out to smoosh us, it could have merely waited where it was and let us bump right into it."[5]

So that was an interesting encounter one dark night between a smart biologist and an apparently determined elephant. Of course, most of the elephant's behavior will forever remain as mysterious as the night itself was. We're not even sure Chadwick faced a single animal. Maybe there were two or three or even more, emerging from the murk at different times.

In spite of all the unknowns, though, Chadwick's description seems to show an animal who did possess a singular kind of focus and determination. It also, and more certainly, demonstrates a set of elephant qualities that are unquestionably true: Elephants are far better hearers and smellers than seers. Their vision is comparatively weak.[6] Their sense of smell is five times more acute than a bloodhound's.[7] Their hearing reaches well below the range of human hearing.[8] Moreover, through biomechanical receptors known as Pacinian corpuscles located in the dense, fatty pads of the bottoms of their feet, elephants also seem to possess a fine-tuned responsiveness to seismic information, such as the subtle vibrations in the ground made by a far-off earthquake, a distant clap of thunder, other elephants approaching from miles away, or—could it be?—a panicked biologist's thump-thump-thumping little feet.[9] That's why a malicious elephant might be such a formidable presence in the dark, and it's one reason why I think Douglas Chadwick's description of the event is interesting.

The other reason I consider the description interesting is what it suggests about human language and thinking. Yes, you are reading a book about animals and animal behavior. But so much about animals and their ways remains vague, mysterious, unknown; and we are left, so often, looking at animals and what they do through a dark glass, seeing only momentarily the beast emerge, as if on a dark night, before disappearing silently, then quietly emerging once more only to disappear in silence once more. Our knowledge is still limited, in other words, and what we do know about animal behavior is powerfully obscured by long-standing habits of thought. Thus, before you and I can begin to talk seriously about animals, we should first consider the nature of our thinking about them.

I could easily find, in almost every book about animals I have in my bookshelf, a chapter, paragraph, or passage that illustrates what I mean. But since I've already mentioned a biologist's popular book that contains a description of his encounter with a big, bad beast, let me revisit Douglas Chadwick's account to make my point. He wrote, "I will never know what that elephant had in mind that night, but upon reflection, I have to credit the animal with giving us fair warning. If it had really been out to smoosh us, it could have merely waited where it was and let us bump right into it."[10]

Probably the passage does not at all strike you as odd. Nor, perhaps, will it seem odd even when I point out the logical contradiction it embodies. Chadwick tells us, in the first part, that the animal in question has a "mind." He implies that the animal made deliberate choices and had emotional responses. Then, in the second part of the passage, he reiterates four times that the animal is an "it," which is the same pronoun we use when referring to a lifeless, mindless, emotionless, brainless, faceless, random bit of inert matter. A thing. So who or what is this creature: an animal with a mind, with emotions and some capacity for deliberation, or an inanimate thing that belongs in the same category as a rock, a stick, a clod of dirt, or a lump of coal?

A dozen other linguistic habits tell a similar tale. Animals are *trained*, humans *taught*. Animals have *fur*, humans *hair*. Animals operate by *instinct*, while people are moved by *plans* and *ideas*. A newborn animal is a *cub* or *puppy* or *calf*, while people come into this world as *babies* and are soon transformed into *children*. An animal might be an *adolescent*, but only a person is a *teenager*. An adult animal will be either *male* or *female*, but never a *man* or *woman*. An animal can be *killed*, but only a person can be *murdered*. A dead animal makes a *carcass*, whereas a dead person becomes a *corpse* or even, under the right circumstances, a *body in repose*. Indeed, animals decay and disappear entirely after death, while only humans, so we tell ourselves, can hope to find some kind of coherent, soulful existence on the other side. Animals *die*. Only you and I are going to *pass over*.

You might argue that humans really do have minds, whereas animals obviously do not, although your primary evidence for such a remarkable exclusivity may be your own conviction that it must be so, and when I ask you to define what a "mind" is, you are hard-pressed to respond. You might insist that humans really will find life after death, although you will be left explaining a strongly held belief that can neither be proven nor disproven. You might say there is an actual difference between *hair* and *fur*. You might want to point out the fine distinctions between *trained* and *taught*—and call my attention to the fact that in some circumstances, such as with the kind of repetitive muscle learning that athletes endure to perfect their skills, we do talk about humans *training* or *being trained*. Our linguistic habits can, in short, be complex, and we rely on creaky old words to make fine shades of meaning concerning the nature of the tangible, observable world. So perhaps the simple pronoun convention—the matter of *it* versus *he* and *she*, as well as *that* versus *who* and *whom*—illustrates my point as well as anything.

Words project thought. The structure and habits of our language are flags, reasonable indicators of the structure and habits of our thinking, including our ordinarily invisible presumptions and prejudices: the distorted lens of our own minds. And in the case of our usual thinking about animals, the common habit of creating one thought island for people, the island of *who* and *whom*, and a second thought island, that of *it* and *that*, to contain that vast world made up of all animals and all things, suggests an astonishing conceptual divide that simply fails to reflect reality. The reality is this: We are far, far more closely related to any animal than we are to any object. And to mammals, that special group of animals who are primarily the focus of this book, we are a good deal more closely related than we ordinarily admit.

Yet still, we tell ourselves, the conceptual divide is there for a reason. Consider how uncomfortable it can feel when people radically violate the separation represented by those two thought islands. Consider the case of elephant executions.

Topsy had been a working elephant in the Forepaugh Circus in Brooklyn, New York, until one day in May of 1902 when a circus hand named Blount staggered too close while holding a glass of whiskey in one hand and a lit cigar or cigarette in the other.[11] Perhaps the potent smell of alcohol set Topsy off. According to some accounts, Blount tried to feed the elephant a lit cigarette. We do know that within a few seconds, Blount was dead, and within a few weeks, Topsy was sold to Luna Park amusement park on Coney Island and placed under the charge of a trainer named Whitney. Whitney himself had a drinking problem. One day that December, Whitney took Topsy for an unauthorized walk, and the elephant spooked a crew of construction workers. As a result of that unfortunate episode, the trainer lost his job, while the trainee was identified as a dangerous beast and condemned to death.

She had killed that circus hand Blount, after all, and now the story emerged that she had also killed two other circus workers in Texas sometime previously. The owners of Luna Park decided that her execution would be death by hanging, and they built a scaffold and gallows, fenced off a portion of the amusement park to keep away the nonpaying riffraff, and designated a seating arena for the paying public to bear witness. They sold tickets, hired a cameraman to record the event on celluloid for posterity, and settled on the execution date of January 4, 1903.

Meantime, the Society for the Prevention of Cruelty to Animals pronounced that hanging an elephant would constitute cruelty to a dumb animal, so the Luna Park owners settled on electrocution. The electrocution was delayed when Topsy refused to climb onto the scaffold, where the apparatus had been set up, and thus, unable to bring the elephant to their equipment, the electricians were forced to bring their equipment to the elephant. The film of the event shows a rising burst of smoke followed by a falling elephant.

Then there was Mary, the largest member of the five-elephant show featured by the itinerant Sparks Circus.[12] After an early performance on September 12, 1916, the circus paraded down the central thoroughfare of the eastern Tennessee town of Kingsport, headed for the town pond where the animals would be able to drink and bathe. The pachyderms paraded in their usual fashion, with Mary at the lead, the four others following single file behind, each grasping with her trunk the tail of the one in front. People had enthusiatically lined up all along the way, eager to see the big animals. During this brief amble on the way to the pond, though, Mary noticed a fresh watermelon rind, and she paused to investigate the object with her sensitive trunk tip.

A newly hired circus hand unwisely placed in the charge of the elephants prodded Mary with a stick. She did not respond. The worker prodded more emphatically. He poked—or perhaps struck—her. Mary now responded. With her trunk, she picked the man up and threw him against a nearby wooden soft-drink stand, then walked over to where he lay in the dirt and with a front foot crushed his head. The people who had gathered to be amused by the big animals now panicked and began running. Mary slowly backed away from the body on the ground. The elephants behind her, excited now, began trumpeting wildly. Then a number of circus workers moved in to calm the elephants, and soon the crowd had turned into a chanting mob: "Kill the elephant!"

Charlie Sparks, the owner of the circus, mollified the mob by repeating the lie that it was not possible to kill an elephant with a gun, and so he was able to bring all five animals safely back into the perimeter of the circus encampment. That night, the workers packed up their tents and animals and equipment, placed everyone and everything back onto the circus train, and rode the clicking rails to their next destination, which was Erwin, Tennessee. Sparks may at first have hoped the episode back in Kingsport would quickly be forgotten. Mary was worth several thousand dollars. She was the featured attraction of the whole show, the dramatic centerpiece for its advertising. Still, Sparks was aware of the rumors that Mary had killed before, and soon he began to hear the rumors that a lynch mob was preparing to drag a Civil War cannon into town to shoot the beast. Sparks decided to execute Mary himself.

After a matinee show in Erwin (which took place without Mary), the five elephants were gathered into their usual parade formation, with Mary at the lead, the four others behind, holding on to one another trunk-to-tail. In that way, the condemned animal was kept calm and paraded out to the railroad yards of Erwin, where there was a hundred-ton, flatbed-mounted railroad derrick. The five-ton elephant was led to the tracks in front of the derrick. A heavy chain was wrapped around her neck and linked to a giant hook at the end of the derrick's boom. Mary's four companions were taken away, brought back to the circus grounds. Then, as a crowd of about three thousand people looked on, a powerful engine started up, a winch began to turn, a chain began drawing tighter, and a now frantic and struggling Mary was soon lifted right off the ground. The chain broke. The elephant dropped back down onto the tracks, breaking her hip. But a stouter chain was soon located, another noose worked around the animal's neck, and this time it held.

A third elephant execution took place in the fall of 1929, not so long after the train carrying the Al Barnes Circus rolled into Corsicana, Texas.[13] There a huge male elephant named Black Diamond knocked a woman down and stabbed her fatally with his tusks. The pachyderm was soon rushed back to protective confinement at the circus grounds, but an angry crowd of vigilantes had to be driven away from the elephant car. Then local newspapers began promoting the call for justice. By the time the circus had reached its next stop, Corpus Christi, a firing squad of twenty men was already organized. Not long after he emerged from the train, Black Diamond was bound in chains, marched out to a field with the assistance of three other elephants, then chained fast to some trees. It took 170 bullets to finish him off.

As a matter of fact, circus and menagerie owners frequently killed out-of-control elephants in Europe and America during the nineteenth and the first half of the twentieth century. Elephants were killed by cannonballs as well as bullets, were drowned and poisoned as well as being electrocuted, hanged, and pitchforked to death. Probably because the males grow to be twice as big as females, and also because adult males experience periodic surges in testosterone (the condition of musth), making them more aggressive and less predictable than usual, from around the middle of the nineteenth to the middle of the twentieth century, male elephants serving the American entertainment business were almost routinely killed. By the start of World War II, only a half dozen mature males remained out of a total of 264 elephants alive in U.S. zoos and circuses.[14]

Most of those killings may not violate our ordinary sense of what's normal or appropriate. People normally put down rabid raccoons, shoot injured horses, euthanize vicious dogs, and so on. It's done for their own good or for the protection of the public, we say. And so we might imagine that the history of all that elephant killing has been essentially an unpleasant sequence of put-downs made ugly only by the size of the problem. Maybe so. But I would distinguish the events I'm calling "elephant executions" from a more ordinary put-down or unavoidable euthanasia in at least three ways.

First, the executions were all responses to elephants killing people in ways that seemed deliberate. The killings may not have been, as we say when humans do them, premeditated, but they were also not mere accidents, the untoward result of a big foot accidentally placed in the wrong spot. Elephants are not clumsy.

Second, the executions were all in some sense collective or social. They were either done in public, before an audience, or at least carried out by a public or with a public in mind. People bought tickets and sat in the grandstand to watch Topsy's electrocution. A crowd of three thousand gathered to see Mary hang. An enraged mob in Corsicana, Texas, tried to break into the circus to take revenge on Black Diamond.

Third, although these executions may have been done with the safety of the public in mind, they also seem to have been motivated by a desire for justice. Or was it vengeance? The word "justice" implies a methodical application of law and custom, so perhaps these events could more rightly be called elephant "lynchings," and perhaps "vengeance" is the more likely motivation. But vengeance, simpler and cruder than justice, still implies moral outrage based on the idea that a guilty party is responsible for having done something terribly wrong. And responsibility, in either case, may be the important commonality here. Among the most telling characteristics of thingness is our certainty that a thing, a stick or stone, is not and never can be responsible. Responsibility requires a psychological presence, a *mind*.

Executing an elephant for the crime of murder strikes us today as profoundly irrational, a blatant sort of anthropomorphism. It seems, in a word, "medieval"—and with good reason.

The medieval concept of the universe was of a place gloriously filled with minds: from the mind of God on high to the minds residing in the planets and planetary spheres, the minds of angels moving across the ether and demons riding through the air, the minds of the men and women on Earth's surface at the center of the universe, the minds of the animals also living there, and finally down to the mind of Satan himself, situated in Earth's fiery interior. This imagined universe cohered in part because all those minds shared some fundamentally humanlike qualities. While it was true that no finite human mind could hope to comprehend the infinite mysteries of God's mind, it was also true that the main goal for humans on Earth was to move closer in their understanding of the divine mind.

This universe operated as a physical entity because of the principle of *kindly enclyning*, which, in a fashion analogous to our own imagined principle of gravity, meant that intelligences of a similar kind were inclined toward each other.[15] The Intelligence of God was kindly enclyned to the intelligences embodied or residing in the invisible, concentric spheres that made up the superstructure of the universe.[16] Through the kindly enclyning of Intelligence for intelligences, the outermost of the universe's invisible spheres was set in motion; and its steady revolutions were transmitted, through descending stages of kindly enclyning, to the invisible spheres situated concentrically within. The first of those interior spheres carried the stars. The others carried each of the outer planets, the sun, the inner planets, and the moon; and they all turned around Earth, located at the very center of it all.

The mystery of how humans could be both physical and intellectual beings was explained by a theory of souls, which was simultaneously a theory of mind that distinguished plants from animals from people. All plants were endowed with vegetable souls, which accounted for growth and reproduction. All animals incorporated vegetable souls within a sensitive soul, giving them sentience. Humans maintained both vegetable and sensitive souls within a rational soul, and by virtue of that rational soul a person could hope to touch the unclouded intelligence of the angels below God. At the same time, of course, possessing a sensitive soul kept humans in communion with the sentience of animals, which was itself a surprisingly rich and flexible kind of intelligence.

In the more practical world of jurisprudence, animals were believed to possess sentience enough to be responsible for their actions, and such a possibly charming or eccentric consideration sometimes had bizarre consequences. A domestic animal accused of killing someone could be brought into court, challenged by a prosecuting attorney, supported by a defense attorney, tried according to the rules of law, and, if convicted, executed by hanging, burning, or being buried alive. In his book *The Criminal Prosecution and Capital Punishment of Animals*, E. P. Evans reviews the long history of medieval animal prosecutions and details numerous cases, such as the trial in 1379 of three sows from the communal herd of Saint-Marcel-le-Jeussey for the murder of a swinekeeper's young son. Since the three sows had been incited in their violence by a number of squealing comrades, the rest of the herd and the swine from a second herd belonging to the local priory were charged as criminal accomplices and also tried, convicted, and prepared for the gallows. At the last minute, the local prior, distressed at the prospect of losing the priory herd, petitioned Philip, Duke of Burgundy, on behalf of the convicted accomplices. Upon the authority of the duke, the three murderous sows were hanged while the remaining swine were pardoned.[17]

Second to murder, the most common crime of domestic animals may have been sex or sodomy with a person, a heinous act for which both parties would face the gallows.[18] Possibly the most crowded execution on this theme was carried out in 1662, in New Haven, Connecticut, when a man named Potter was hanged alongside his eight guilty partners: one cow, two heifers, two sows, and three sheep.

Smaller animals committing lesser offenses not requiring the death penalty were sometimes tried by ecclesiastical rather than criminal courts, and the ecclesiastical courts often recognized another interesting thing about animal and human minds: While all minds were fundamentally alike or had overlapping commonalities, they were also essentially permeable. In other words, one mind could be penetrated or taken over by another. And in the same way that people's minds might be possessed by the perverse minds of demons, so demons might also work their evil ways by taking over animal minds. Animals found guilty of bad behavior in an ecclesiastical court might, therefore, actually require spiritual intervention rather than physical punishment. The animals tried in ecclesiastical courts ranged from mice, rats, and birds to insects, slugs, and caterpillars, and their most common offense was destruction of crops. The creatures might still be represented by a defense attorney, but if their case was lost, they could find themselves on the receiving end of a holy anathema. Such a fate was endured by the noxious insects accused of ruining farmers' crops in 1478 and tried before the bishop of Lausanne, Switzerland. Directly addressing the offending creatures from the elevated platform of his pulpit one Sunday, Father Bernhard Schmid, the parish priest of Berne, eloquently and elaborately summoned them or their advocate to appear six days hence, at a particular time and place, to answer for their conduct, whereupon (so Father Schmid warned the bugs ominously) the bishop of Lausanne or his subordinate would "proceed against you according to the rules of justice with curses and other exorcisms, as is proper in such cases in accordance with legal form and established procedure."[19]

The medieval vision of animal minds as intelligent entities constructed in a humanoid form—essentially underendowed human minds—is what I call the First Way of thinking about animals. I like to contrast it with a system of thinking and belief that generally replaced it. This Second Way, which might alternatively be described as the Enlightenment vision, is sometimes associated with the writings of French philosopher René Descartes.

In his *Discourse on Method* (1637), Descartes described a world that has been largely stripped of intelligence and emptied of minds, save for those belonging to God and humankind. Descartes argued that mind and body are strictly distinct entities, and that, since only humans have been given immortal souls, only humans have minds. Animals are still alive and even capable of experiencing sensation, he declared, but rather than possessing humanlike minds, they have nothing of the sort. Animals might appear as if they are intelligent agents, but at best one can only say that they have brains and sensations associated with them. Animals, Descartes wrote, "have no reason at all," and as a result "nature acts in them according to the disposition of their organs, just as a clock, which is only composed of wheels and weights, is able to tell the hours and measure the time."[20] Animals are machines made by nature, in other words, and since they are machines, one need not concern oneself when, for example, an animal cries out as if in pain. The animal is not actually in pain, since pain is a mental experience. An animal has no mind with which to register such a mental experience, and so the machine cloaked with the skin of an animal is merely running through some meaningless actions that people falsely interpret as pain.

Herman Melville's classic American novel *Moby-Dick* draws the problem of how to think about animals into the world of nineteenth-century industrial whaling by posing the following dilemma: What if a whale is not an object or a thing, not merely a crop to be harvested for the natural resources of meat and oil, but rather a living biological entity self-directed by some force or quality we might recognize as a *mind*? That radical question drives the springs and gears of this great big book about a big, bad whale. All the characters are swept into circumstances that force them to confront the question, but the two most legally responsible men aboard the whaling ship where the bulk of the action occurs, the captain and the first mate, take the two most clearly defined positions on the matter.

The captain of the ship is an obsessive sort named Ahab, who, having lost his leg to the gnashing teeth of an enormous albino whale named Moby Dick, vows revenge. Ahab, thinking of animals in the First Way, is convinced that the particular white whale he seeks is alive, aware, and morally responsible. Moby Dick, in Ahab's view, has committed an extreme personal violation that requires an extreme and personalized response. First mate Starbuck, who has signed on to the voyage with the commonplace purpose of earning money through the practical activity of harvesting whales, embodies Second Way thinking about animals.

Since it soon becomes clear that Captain Ahab's monomaniacal pursuit of a single whale threatens not only the economic success of the voyage but the very lives of the crew, the first mate is provoked to challenge his captain. "Vengeance on a dumb brute!" Starbuck cries out in alarm. To take revenge against a mere animal, a "dumb thing" that acted from "blindest instinct," is a terrible error. Starbuck is portrayed as a physically courageous man, and his words here challenge not so much Ahab's dangerous plan of action as they do his alarming set of ideas. Starbuck, in fact, embraces the values of a conventional man from his time and place, and among his unshakable certainties is that animals are things, that all of nonhuman nature is unconscious and fully disconnected from the psychological reality known only to humans. To think or to behave otherwise, he blurts out to Captain Ahab, is to violate the most fundamental tenets of secular and religious convention. It would be madness and blasphemy.

I believe that both of Melville's characters, both Ahab and Starbuck, are speaking and acting in error, and that the deepest truths about animals are to be found elsewhere. One can take revenge only on a kindred being, and Ahab recognizes correctly that Moby Dick is indeed biological kin: an animal with a psychological presence. Ahab is wrong, however, in imagining this kindred being has a humanlike mind and therefore is morally responsible, in human terms, for the crime of gnawing off Ahab's leg. We have a word for that distorted perception, the overhumanizing of nonhuman animals: *anthropomorphism.*

Starbuck's position, the Second Way, is still commonly accepted. You may believe he's right. But if you believe, as I do, that he presents another kind of false thinking about animals, we still have no quick way to describe it. Starbuck sees animals as things, or as biological machines operating through rigid programs of instinct and reflex, and he thus presupposes a radical division between human and animal. This Second Way perspective confirms the contemporary pronoun convention: animals belonging with the world of objects on the island of *it* and *that.*[21] And it is still the default position for much conventional thinking about animals today. Indeed, the error of this perspective remains common enough as to be essentially invisible. We don't even have a word for it. So let me invent that word now. Let me suggest that Starbuck's error might be called *anthropo-exemptionalism*. We all know that people are exceptional. Of course they are, just as elephants and apes and lions are, in their own particular ways, also exceptional. But the error I refer to is not about thinking of humans and human behaviors as exceptional. Rather, it's the error of considering them exempt: utterly disconnected from the limits, systems, structures, and truths of the rest of the natural world.

In sum, Ahab and Starbuck take radically opposing positions on the problem of a strangely persistent and obviously dangerous animal, and both are wrong. Neither one solves the problem or even, quite, pauses long enough in the action to recognize that there is one. Nor is anyone else on the ship, not even the thoughtful and imaginative narrator, Ishmael, capable of understanding with full clarity the meaning of a mindful whale. A clearer vision is hinted at in some of the language and a few important passages, but this book is wrapped in ambiguity, and I think it's a mistake to hope that the wise author will somehow, at some special moment, lean over and whisper the real truth in our ears.

Nor is the way to proceed a matter of averaging Ahab's and Starbuck's positions, or by allowing for some sort of mutual cancellation. Instead, we should triangulate from them and find a possible Third Way. This Third Way would incorporate Ahab's unshakable conviction that an animal can have a mind with Starbuck's opposing certainty that no one will find a mind out there except the one inside another human. The Third Way, in short, takes a bit of both. It allows for the existence of animal minds, but it considers them alien minds—alien, that is, from human minds. The Third Way looks for both real similarity, between human and animal minds, and genuine dissimilarity.

Herman Melville's famous contemporary and fellow South Seas adventurer Charles Darwin suggested the Third Way of thinking about animals when he promoted his theory of evolution through natural selection in *On the Origin of Species* (1859).[22] Darwin tells us that the enormous number of anatomical similarities we can identify among separate species is the result of a shared evolutionary history. You and I have eyes that are similar to the eyes of, say, dogs, elephants, rattlesnakes, and ring-tailed lemurs, along with a few million other species, primarily because eyes evolved a long time ago, long before those species branched out and developed their particular dissimilarities from one another. A shared evolutionary history can explain much about the vast number of physical similarities among biological entities. Let's call that *the principle of anatomical continuity.* Yet Darwin also understood well that there was no particular reason why such continuity should be limited to anatomy. He argued forcefully in *The Expression of the Emotions in Man and Animals* (1872) for the existence of an emotional and psychological continuity among animals and humans as well, based on a shared evolutionary history, and he also believed in a mental continuity to the degree that one might even speak of animals as having morality or something similar to it.[23] "Besides love and sympathy," he wrote in *The Descent of Man* (1871), "animals exhibit other qualities connected with the social instincts, which in us would be called moral."[24]

But while Darwin's general theory of evolution was accepted by the scientific community during his lifetime, that acceptance was largely limited to an appreciation of evolution as producing anatomical continuity. Emotional, psychological, and mental continuity were far more challenging concepts.[25] They are more disturbing, harder to understand, study, demonstrate, or even talk about, and so the Second Way of thinking about animals—René Descartes's vision of animals as mindless machines—remained the predominant paradigm for another century.

Although you might consider the idea of animal-as-machine to be strangely exotic or hopelessly antique, Second Way thinking has actually become lodged in much of the language we still use about animals. And once we recognize that a machine can be constructed of soft materials and operate through electrochemically based neural circuits, we can see how Second Way thinking adapted itself to some of the demands of early modern science. Mice, rats, pigeons, and other relatively common and easily caged animals could contribute to the science of the human mind, psychology, not because animals had minds but because early psychology chose to ignore the difficult issue of minds altogether and preferred instead to concentrate on understanding behavior. Animals might not think, but they could still behave. While the fashion known as behaviorism dominated psychology until the 1950s, the scientific discipline that directly studied animals in their wild habitats, ethology, focused on the small and relatively simple creatures of Europe: insects, fish, and birds. Until the early 1960s, ethology was almost wholly oriented to describing the typical behaviors of such species, and it moved to understand those behaviors by reference to a series of machinelike systems, such as instincts and reflexes.[26] "Animal behavior was treated as a set of tropisms and taxes," biologist Donald Griffin writes in *The Question of Animal Awareness: Evolutionary Continuity of Mental Experience* (1976), while "a sort of simplicity filter shielded us from worrying about possible complexities."[27]

The Third Way of thinking about animal minds recalls the medieval concept of a world filled with minds, human and animal, except that in this version the minds are impermeable and, in part, alien. *Impermeable* means that they can't be possessed by other minds. *Alien* means that they are not simply smaller or truncated versions of human minds. Instead, they are differentially characterized by fine-tuned evolutionary adaptations to the demands of their species' particular social and ecological niches. Instead of speaking about "the animal mind," as if it were one single entity that can be described simply, therefore, the Third Way recognizes that each species has evolved to live with a different neurological structure, different brain, different subjective experience, different emotions and psychology, different mind. And once we recognize that all animal minds are, from our own perspective, alien minds, we also recognize that they are alien from each other. Don't tell me that, because you believe you understand the workings of a cat's or a cow's mind, you now are ready to understand the mind of a whale. There are animal minds, not an animal mind.

Philosopher Thomas Nagel wrote about this alien quality of animal minds in his 1974 essay "What Is It Like to Be a Bat?" Nagel begins the essay by asserting that "conscious experience is a widespread phenomenon," common to many different animal species. But when we recognize that an animal is conscious, we are then recognizing that the animal—for example, a bat—has a subjective mental experience, which means that we can legitimately ask, "What is it like to be a bat?" We can ask the question, but we can never fully answer it because of the inherent limitations of our own minds. We are not bats, and the structure of our nonbat minds makes it impossible to know the subjective experience of a bat. We could try to imagine having wings made of skin stretched between our arms and torsos. We could speculate about the experience of having bad eyesight but hearing acute enough that we can navigate in a dark room by listening to the echoes of our own squeaks. We could think of ourselves as driven by a compelling appetite for insects and as possessing a penchant for hanging upside down in a musty cave during the day. But imagining these various capacities and inclinations, Nagel declares, will never take us beyond the limits of our own human mental experience. To know perfectly what it is like to be a bat would require that a person have a bat's mind, which is not possible.[28]

It's the same with dogs. When I walk my dogs, Smoke and Spike, two sisters of Saint Bernard maternity and obscure paternity, who together weigh approximately what I weigh, my inclination is to walk in straight lines, whereas theirs is to ambulate in crooked ones. I can understand the need to relieve oneself from time to time, but why in public without even an ounce of modesty? And is all that sniffing of fire hydrants really necessary? I could probably figure out the logic of the sniffing. Something to do with territorial concerns or the eternal search for friend and foe, food and sex. But at best my explanations will be objective assessments of a subjective mental experience that I may never fully grasp because it's alien to my own experience. I not only lack the nose, but I also lack the neural circuitry that draws an olfactory experience into the very core of my being. My nose contains around six million olfactory receptor cells, which seems sufficient—but a beagle has fifty times that number.[29] Because a dog also has more kinds of olfactory cells acting in different combinations, it could be that any real comparison will conclude that the beagle is actually millions of times more sensitive than I to the world of floating molecules that may be perceived as smells. To use a comparison suggested by psychologist Alexandra Horowitz in her book *Inside of a Dog* (2009), you and I might be able to detect a teaspoon of sugar in a cup of coffee, but a dog can detect the same sugar dissolved in two Olympic-size swimming pools.[30] In short, I will never know experientially what it's like to be a smelling dog in a smelly world. To a significant degree, it's just an alien psychological experience. Having been around dogs for much of my life, however, I seldom think about their alien qualities, and my current relationship with those two sisters, Smoke and Spike, is based on a quiet compromise. Much of their inner world is alien from mine, much of mine for theirs, and so we are left with a relationship in those parts of our worlds that do overlap. We are friends, my dogs and I, in spite of evolutionary discontinuity and because of evolutionary continuity.

Sure, I can see my two dogs through the distorted lens of anthropoexemptionalism. I can imagine, wrongly, that we have nothing in common. I can pretend that I am a mindful creature of high intelligence, whereas they are two mindless bundles of meat and bone rudely governed by a machinery of impulse and instinct. I can fatuously declare that "I think therefore

I am"—and that "They do not think, therefore they are not." These are all overprojections of evolutionary discontinuity, the errors of Second Way thinking about animals, and in the case of our beloved domestic pets, they fail to account for one otherwise strange phenomenon: We care deeply about them.

Why do we bother giving our dogs names, stroking them fondly on the quivering canine cranium, talking to them, allowing them to climb into the bed on cold and lonely nights, and responding sympathetically when they make joyful motions with their tails or sad expressions with their faces? If we were truly exempt from the natural world, we could easily replace our dogs with fur-covered robots and never again have to clean up the mess. We care about our companion animals because we share with them some essential sense of the world and some basic emotions that govern our feelings about the world. We talk to our dogs because we know they listen, and if their understanding is limited to a few basic words and concepts, so much the better. We also love them for their powerful simplicity.

Our connection with other species oscillates somewhere between discontinuity and continuity, while our errors in thinking about them likewise oscillate between false anthropomorphism (an exaggerated assertion of continuity) and false anthropo-exemptionalism (an exaggerated insistence on discontinuity). The Third Way recognizes this fact. It appreciates that the danger of false thinking about animals is two-sided.

Everyone who owns a dog or cat knows intuitively that these animals have minds, but dog and cat owners also recognize that their favorite animals' minds are in some ways limited, at least from the human perspective. After all, our beloved pets, clever and sweet and wonderful as they may be, still need us to open doors, locate the food, and drive them to the veterinarian when they're ill. The issue of animal intelligence—how smart any particular species is, or in what forms their smartness appears—is not the subject of this book. Rather, this book is my attempt to trace evolutionary continuity in the area of moral psychology. It is not an attempt to invent continuity where it does not exist, nor am I hoping to make animals look smarter or nicer than they actually are. I use the word *animals* here primarily as a convenient shorthand for that group of warm-blooded and relatively large-brained species known as *mammals*, and this book describes a moral psychology that is in some ways distinctively mammalian and, in most ways, based on emotional rather than intellectual structures. Just as it seems self-evidently true, within our own species, that how smart you are is generally unrelated to how good you are, so I will assert that morality among animals is only grossly related to any particular species' level of cognition or intelligence.

My argument divides into four parts, with Part I promoting the understanding of morality, both human and nonhuman, as a gift of biological evolution. This is the section on *where morality comes from*, and it requires presenting a few basic concepts, including that of our linguistic and intellectual biases (as described in chapter 1), the orientation problem (chapter 2), a working definition of morality (chapter 3), and a theory of morality's structure (chapter 4).

Since I conceive of morality as a complex system with two dynamically interacting parts—rules and attachments—my description of *what morality is* also requires two parts. Part II of this book, then, is devoted to the rules of morality. For humans, these rules have been written down, but they evolved long before language and for nonlinguistic reasons. The rules of morality evolved in response to social conflict, and we can predict what those rules will look like by identifying those realms of social life where conflict is likely to be most intense. I identify five such realms, which are the subjects of chapters 5 through 9: authority, violence, sex, possession, and communication. For language-using humans, we can locate the rules covering these five realms in the written text of various moral codes, such as the Ten Commandments. For non-language-using nonhuman species, we are forced to discover the rules mainly by looking at behaviors, including antisocial behaviors that challenge the rules and pro-social behaviors that confirm or enforce them.

In Part III, I introduce attachments morality, which includes the mechanisms promoting cooperation (described in chapter 10) and kindness (chapter 11). These attachments are fundamental principles, more general and pervasive than the moral rules. For humans, such abiding principles can be found in the texts of many moral documents, and I refer to well-known passages taken from the biblical New Testament to make this point. But for both humans and nonhumans, we understand attachments as part of the urge to transcend selfishness in ways both practical (cooperation) and impractical (kindness or altruism based on empathy).

Finally, in Part IV, I summarize, synthesize, and speculate in order to consider *where morality is going*. Rules morality and attachments morality look and feel different in part because they emerge from different evolutionary processes—and yet they interact. Both are complex to begin with, but rules and attachments combine and intertwine to create a highly dynamic system that is expressed as the potentially flexible thing we call *morality.* The dual nature of human and animal morality is the subject of chapter 12, and the resulting dynamic flexibility is described in chapter 13. As I note in these two chapters, moreover, the interaction of rules and attachments has the capacity to produce even greater levels of dynamic flexibility among those several species, including our own, that show average gender differences in how the rules and attachments are experienced and expressed.

Such ideas are speculative, obviously, and my final chapter, chapter 14, may be the most speculative of all. Here I argue that the evolutionary and cultural future of human morality could bring an increased importance for empathy and, with it, an expanded role for attachments morality. I express the hope that such changes might, in the distant future, lead us in the direction of greater tolerance, higher wisdom, and a new condition of peace between humans and nonhumans alike.

CHAPTER 2

Orientations

And still deeper the meaning of that story of Narcissus,
who because he could not grasp the tormenting, mild image he saw
in the fountain, plunged into it and was drowned. But that same
image, we ourselves see in all rivers and oceans. It is the image of the
ungraspable phantom of life; and this is the key to it all.

—Herman Melville, *Moby-Dick*[1]

I first became interested in the problem of animals about twenty-five years ago. At the time, I imagined "the problem of animals" to be the problem of animals going extinct, and I set out to learn why.

I had been a literature student, a scholar and reader of great books, but if literature is a fair reflection of the human mind, then animals are just about absent from it. In any case, I had already spent so much time navigating the gloomy labyrinths of formal education that I was finally handed a Ph.D. and shown the door. So that was enough, and I did not take this new interest in a new subject back to old places. Rather I took it in a delightfully contrary direction: to the sunny outdoors, first to the woods of southeastern Brazil and the Brazilian Amazon and then down the Amazon River in a small boat, where I swung lazily back and forth in a hammock, puffed sybaritically on a big cigar, and considered methodically the map spread open on my lap.

I was on a quest. Financed by a pocketful of credit cards, I was traveling around the world. First stop: South America. Then it would be on to Africa, Madagascar, India, and various spots in Indonesia and Malaysian Borneo. I intended during my extended circumnavigation to find the world's dozen rarest and most endangered primate types and learn at first hand why they were going extinct.

It was the sort of thing bird-watchers do. I had put together my own master list of rare and endangered primates. I had already found individuals from two of the most highly endangered monkey species in southeastern Brazil, watched them, took some snapshots with my little camera, jotted down impressions in my little notebook, and checked those two off the list. Now here I was looking for number three. I was hoping to sight individuals from a type of beautiful, black, puffy-tailed, and large-bearded monkey known as the southern bearded saki. I also intended to consider these elusive creatures in context, to learn more generally about why they were so endangered. These were the most endangered monkeys of the entire Amazon Basin, an almost unconceivably vast and diverse region that contains numerous different monkey types, of which about half were in trouble. But why? And why was this southern bearded saki in the worst shape of all? The answer to those questions, I knew, would at least partly be connected with knowing why and how the Amazon forest itself was under serious threat.

Such larger issues had brought me, in a brief detour, to the geographical navel of the Amazon Basin, the seedy river-port city of Manaus, where I located the offices of a group of biologists working on something called the Minimum Critical Size Project. I had already written to someone in this group and gotten myself invited on one of their expeditions into the rain forest, where they were conducting various individual research projects while monitoring the rates of biodiversity disintegration—in plain English, *species loss*—occurring when certain-size sections of rain forest are turned into ecological islands surrounded by seas of burned-out and cleared-over cattle ranch.[2]

Soon I was bouncing in a Jeep across muddy roads and through the moonscape of cattle ranches—rough grass and stark sunlight, some languid cattle, a few industrious cowboys on horses, ten thousand dead trees, ten billion excited termites—out to the great Amazon rain forest. Once inside the forest, we set up camp on a bit of higher ground surrounded by a wealth of flora and fauna. This little expedition included me, an entomologist, an ornithologist, and two worker-porter-cook-assistants, and our camp consisted primarily of hammocks strung between poles beneath a plastic roof. During the day, the entomologist occupied himself with chasing termites, while the ornithologist was hoping to study the feeding behavior of a particular bird. First, though, she wanted to understand the fruiting cycle of the bird's favorite fruit, which appeared high up in a big tree known as *Clusia grandiflora*, so her days were spent chasing *Clusia grandiflora*.[3]

Rita Mesquita was her name. She was a small, attractive young woman with arched dark eyebrows and black hair kept under partial control with a lavender loop of twine. She spoke Portuguese in quick, dramatic, often hilarious bursts to her assistant, and English in slow, staid, semigrammatical productions for my benefit, and I followed her and the assistant around as they walked from one *Clusia grandiflora* to the next. She had marked out a whole complicated maze of routes to reach thirty-two such trees. She had a compass and a plastic-covered map of the routes to keep from getting lost, and thus she would check the compass, consult the map, then walk fast to the next *Clusia grandiflora*. Stop directly beneath the tree. Pull out her binoculars. Bend her neck way, way back and look almost straight up into the top of the tree, where the fruits were revealing themselves in various stages of ripeness. She took notes, gathered and bagged fruit samples from the ground, considered her compass and map, and then we were off to find the next *Clusia grandiflora*.

So it was in the company of a beautiful Brazilian ornithologist that I had my first experience walking and camping inside the Amazon rain forest, and I loved it. The place was stunning: giant trees and a high canopy of shivering leaves, bursting plants, and obscure animals; knotted vines, dropping lianas, and the occasional beautiful bright flower appearing where you least expected it; a vast dynamic of light and space and glowing color. Those were the glories of the forest. The dangers I more slowly came to recognize—starting when I made my first solo foray out of camp in order to answer the urges of nature and modesty simultaneously. I walked away from my companions for about forty paces, reaching a place where I could no longer see them, or they me, and a few steps later I found a spot where I was suddenly no longer certain where I had just come from. I had begun to lose my orientation, and thus I became concerned that I would soon be lost, followed by dead.

The sun was directly overhead. The high, dense, and complex canopy blocked most direct sunlight in any case. Even if I could have recognized this or that species of tree, I would never remember individual trees. The terrain was monotonously rolling, an endless roller coaster of ups and downs with no signature irregularities. Nor were there any orientation clues down on the forest floor; and now, looking up and looking around, I felt like an insect trapped inside a big bottle. I may have wanted to keep on walking, but did I have any idea where I was headed? Rita Mesquita had marked out her trails connecting the *Clusia grandifloria* trees. She routinely referred to her map of the trails. She regularly consulted her compass. I could only refer to and consult with her.

I called out. No answer. I called again. She called back. In that way, I soon found my way back to camp. But I had just learned my first critical lesson about the Amazon rain forest: *You do not go into such a place alone.* The problem is not poisonous snakes, ferocious carnivores, pernicious pathogens, or any other kind of aggressive biological entity. The problem is orientation.

One essential quality of a tropical rain forest is diversity, an intensely concentrated crowding of species. On average, a four-square-mile piece of tropical rain forest will contain 100 species of reptiles, 125 species of mammals, 150 of butterflies, 400 of birds, and roughly 42,000 of insects. But the flip side of diversity is dispersion. You might find a hundred different tree species in a two-and-a-half-acre patch of the Amazon, but walk half a mile distant, examine a comparable-size patch, and you'll find that half the tree species are new.[4] The principle of dispersion works on a larger scale, too. The Amazon forest, for example, might actually be considered as eight different forests, eight distinctive assemblages of plant and animal types.[5]

So species are intensely plentiful but spottily distributed in a tropical forest. In terms of monkeys, the full Amazon is home to a large number of species. Let's say that number is thirty. But you'll never find all thirty in one area. One species will inhabit a comparatively small triangle of swamp forest six hundred miles in that direction, another may be living along an elongated strip of riverine forest right in front of you. A third will be found in a somewhat larger patch of terrestrial forest a few hundred miles over in that direction.

Here's where the boat, the hammock, the cigar, and the map spread open on my lap come in.

The boat was necessary because the monkeys I was looking for, the southern bearded sakis, lived in a patch of forest about 750 miles downriver from Manaus. Boats in the Amazon Basin are like buses elsewhere, so you might imagine I had purchased a ticket for the Greyhound and was taking a cross-country trip in the USA.

The hammock was essential for sitting and sleeping. I was traveling on the open lower deck of this vessel, lined up alongside maybe two dozen other people, all local Brazilians, with another couple dozen lined up in their hammocks on the other side. I strung my own hammock between a pair of hooks established for just that purpose, wrapped myself in it at night and snoozed like a butterfly in a cocoon, and, during the day, relaxed more openly on it. From that vantage point, I could watch the brown and churning water move in one direction and the green and hazy shore pass by in the other.

The cigar? This open-air vessel had constant crosswinds, and I was still insensitive enough to smoke around others and immature enough not to think deeply about the relationship between tobacco and life expectancy. Also, while my Spanish was only minimally functional in a Portuguese-speaking country, I found that passing out cigars to some of my fellow passengers was its own sort of friendly conversation.

The map spread open on my lap would help with orientation—although, actually, I carried several maps of three different sorts. First, I had a star map of the southern hemisphere, given to me by an astrophysicist friend from the northern hemisphere. The star map and the stars above, sparkling like white diamonds in black velvet, provided me with an important kind of macrogeographical orientation—as I looked at a piece of the universe from a new perspective and watched, for example, the Southern Cross rise up toward the middle of the night's black dome.

Later on, I began to see that the star map and the stars also provided an orientation in time. Now that I had embarked on my global monkey chase, I wanted to consider time in the big way, the way evolutionary biologists do, and I imagined that the stars might help. We commonly believe that the stars are utterly unchanging, permanently fixed in their positions—but of course they are not. As the universe expands, the hurtling stars within it are slowly shifting their positions in relation to one another. Edmond Halley, remembered as the man who discovered a big comet, was the first astronomer to note that even since the time Ptolemy had carefully charted his own star map, a mere fourteen hundred years earlier, some of those celestial bodies had changed position. In a similar fashion, the stars making up the Big Dipper will have shifted their positions enough that within another two hundred thousand years, more or less, that constellation will have to be renamed. Let me be the first to suggest it: Big Frying Pan.

The constellation Orion the Hunter can be identified partly by the faded red star marking his right shoulder, Betelgeuse, an ancient, dying star: red because it has become comparatively cool, large because its gravity has become relatively weak. Betelgeuse is seventy-five million years old, but most of the other stars making up this constellation are white-hot whippersnappers, including the white diamond Rigel, marking Orion's left foot. Rigel may be only two to three dozen million years old.

Thinking like an evolutionary biologist, as I was trying to do, requires orienting oneself within that sort of time frame. The ecosystem we now describe as the great Amazon rain forest began around seventy-five million years ago, meaning that the forest I was then sliding past was a contemporary of Betelgeuse's. The primates I was looking for began as a scurrying, furtive group of mammals around sixty-five million years ago, long before the birth of Rigel—about the time that a giant asteroid struck the Earth, exploded, and raised a dark cloud. With the sunlight shuttered for a season or two, that dark cloud pretty much wiped the slate clean: extinguishing the last of the big dinosaurs, for example, and enabling some small mammals to begin flourishing in newly attractive environments in the trees. Those became the first primates.[6]

I find thinking with that sort of time orientation reassuring in some ways but distressing in others. After all, the changes we ordinarily conceptualize as happening somewhat gradually—such as the tenfold expansion in human numbers since, say, the publication of Herman Melville's big whale book, *Moby-Dick* (1851), and the appearance of a billion gasoline-burning, carbon-dioxide-emitting automobiles in our own era—are from a biologist's perspective occurring with the speed and violence of an exploding asteroid. Politicians, who ordinarily look through a time window about four years wide, should speak more often to biologists.

The other two kinds of maps I carried and would sometimes spread out on my lap were useful for geographical orientation. First there was my big map of South America, good for tracing the vermiculate curves of the Amazon and identifying a few of the more salient villages and towns I was passing and the big city way down at the mouth of the river, Belém, that I was bound for.

In Belém, my final map came into its own. This was something entirely different from my big South American map, with its childishly bright colors and dot-matrix splatter of cities and towns and sometimes villages, its squiggling rivers and the mostly straight lines demarcating national borders. This was a much smaller and more subdued map of another world with an alternative set of national boundaries defining the territory of the southern bearded sakis. It was a primate-distribution map, one of several I had photocopied from a wonderful book, *Primates of the World* (1983), put together by Jaclyn Wolfheim.[7] Wolfheim's encyclopedic work identified all the world's primates (monkeys, apes, and prosimians—such as lemurs) and described where they were, what they did, how they lived, what and whom they ate, who ate them, how their societies were structured, and so on. The book also provided a distribution map for each primate type that marked out, while identifying standard rivers and other markers, the species' territorial boundaries. I was fascinated by Wolfheim's distribution maps and carried several of them with me on this trip. And now, having arrived in Belém and rested overnight in a cheap hotel with peepholes neatly drilled into the doors and walls, I wandered through the city until I had located South America's most important tropical botany, zoology, and anthropology research institute, the Museu Goeldi. There I found an American-born ornithologist named David Oren and showed him my distribution map of the southern bearded saki.

Oren studied the map. He showed me some mothballed skins and skulls of these monkeys kept in drawers at the museum. And, after some discussion and more consultation with more maps, he suggested that I might find still-living monkeys if I traveled west and then south from Belém, up a tributary of the Amazon—the Tocantins River—for a day and a half until I reached the old village of Tucuruí. On the far side of that old village I would find a gigantic hydroelectric dam, and on the other side of the dam I would find nine hundred square miles of a newly created reservoir. The reservoir was sitting on top of what used to be nine hundred square miles of virgin rain forest, territory of the southern bearded saki. Oren thought I might be able to find some of the remaining monkeys in forests around the edge of the reservoir, and he arranged an introduction with some people associated with the company, Electronorte, responsible for the dam.

Within a few days, I had reached the village of Tucuruí and was racing in a speedboat across the bright surface of the reservoir, accompanied by an English-speaking biologist, Marion Meyer. The boat took us out to an environmental monitoring station at the edge of the reservoir, where Meyer introduced me to a former hunter of bearded sakis named Pedro Pimentel. In the company of biologist Meyer and former hunter Pimentel, after a long morning's wander through the forest's meandering maze, I heard some high-pitched whistles and squeaks and saw a series of distinctive perturbations in the leaves above. We spent the next couple of days watching and sometimes following a large group of dark, thick-tailed monkeys approximately the size of domestic cats, with humanlike grasping hands, and humanoid faces, with strange upraised knobs of fur on their foreheads and peculiar Brillo-pad beards of fur at their chins, who ate leaves and nuts and leaped from one quivering green mattress to another.

So I checked off the third primate on my list, a beautiful domestic-cat-size creature on the edge of disappearing forever. The Brazilians call this monkey *cuxiu*.[8]

From there, I flew across an ocean and dropped down onto the edge of another continent, where I went through a similar process: looking for a primate needle in the tropical-forest haystack. I landed in the small West African nation of Sierra Leone and persuaded someone from the Peace Corps office in Freetown, the capital city, to give me a two-day ride in the cab of a small, breaking-down truck. On the evening of the second day, I arrived at the tiny, moonlit village of Kambama. After some candlelit discussions with the village chief, I was ferried in a rubber dinghy onto the Moa River, past the resident crocodile, and out to Tiwai Island, which for historical reasons (several decades of hunting prohibitions maintained by the local chiefdoms) was filled with monkeys, including the species I sought.

As ever, my big map—the political map—provided enough guidance for me to get from one part of the human world (a nation, a city, roads and towns and villages) to another part, while my little maps—the distribution maps—enabled me to travel from one primate world to another. Primate worlds: I had begun to think that way. And I had begun to see my distribution maps as showing a series of secret doorways into those hidden worlds. Every trip makes the traveler look at maps a little differently, and I was no exception. Now, when I see a big map entitled "South America," I say quietly to myself, *Yes, that's South America, with all the fingerprints of the human political and demographic presence—and where are the animals?* And when I see a big map that identifies itself grandly as "The World," I think to myself, *This is really the human political world. But over there is where the Orangutans live. There is the land of the Mountain Gorillas. And here is the besieged little world of the Golden Lion Tamarins.*

Primate worlds. Not very scientific, you might be thinking. But here's the thing. When you recognize that primates, like many other animals and probably all other mammals, have a subjective mental experience of the world, then you are saying that they have a psychological presence.

They have minds. They see, hear, feel, remember, and even possibly calculate and plan. They may have intentions. They are aware, and they could have some level of self-awareness. They might have some sense of what others around them are intending, planning, or thinking. They will most likely recognize each other as individuals with individual personalities. They will gather into groups in which each individual appears to maintain a richly detailed attitude toward every other individual based on their shared history, and in which all are engaged in a present social network consisting of a highly complex series of relationship negotiations.

Primatologist Susan Perry addresses this idea in the book *Manipulative Monkeys* (2008), which is her account of fifteen years' research on white-faced capuchin monkeys living in the Lomas Barbudal forest of Costa Rica. When tourists and local visitors come to the forest to look at animals, she writes, she finds herself inevitably "struck by the profound difference in the way they perceive the monkeys, compared with my own perception." These nonscientific visitors seem to think of the monkeys as simple clones of each other. The visitors can see some size differences and make out the sex difference, but aside from those standard distinctions, the monkeys all look the same. "Most of the tourists and locals," Perry continues, "think that after five minutes of observing the monkeys, they have learned all there is to learn about these animals: they are black and white, furry, and live in trees."[9]

When Perry looks out over the forests of Lomas Barbudal, she understands them to be seething with simian activity. Out there, in the shadows, behind those leaves, beneath that underbrush, are female white-faced capuchins "feverishly grooming the same female relatives with whom they have been allied for many years, while other group members tend to their infants." Juveniles will be rushing about, "wrestling with one another and perhaps inventing new games that will prepare them for the challenge of coalitionary politics once they are adults." The adult males, meanwhile, might seem to be remote and unengaged, sitting out there on the edges of the social scene, but they are "nervously eyeing allies for signs of treachery and keeping watch on opportunities to better their social position." These males will travel through the forest hoping to develop an important alliance with another male or to mate with a receptive female; but they always recognize that the "reception they receive from any particular individual can vary from warmest affiliation to lethal aggression, depending on who else is present when this occurs." Such is the ordinary drama of social life for these intelligent animals, and it requires "their cognitive machinery [to be] constantly whirring as they try to remember who is friends with whom and under what circumstances."[10]

That's a primate world.

By the end of my trip, I had solved what I had originally considered "the problem of animals," at least for those particular animals. Why were they going extinct? I had the answers. The first primate species I found in southeastern Brazil—large, prehensile-tailed high-wire artists known as muriquis—were survivors of an industrial and urban expansion in Brazil that had already destroyed around 98 percent of the forests they lived in. Because muriquis are also distinctly large—the largest monkeys of South America—they have made good targets for hunters, too. The combination of habitat loss and hunting had, by the time I found them, cut their numbers down to around 350 to 400 individual animals altogether, which represents significantly less than 0.1 percent of their estimated pre-Columbian population.[11]

The second monkeys I located, golden lion tamarins, are arguably the most beautiful animals in South America, surviving, by the time I showed up, in the remaining 1 percent of their original habitat, which consisted of about ten tiny, degraded, and discontinuous pieces of forest. Although habitat loss has always been the major threat to these monkeys, their exceptional beauty also at one time made them favorites for live trapping and sale on the international market as pets and zoo specimens. When I saw them, their best piece of remaining habitat held around 175 individuals altogether. Their total wild population is probably lower than the total number, a few hundred, living in North American zoos.[12]

The southern bearded sakis, my third species, inhabits a piece of the Amazon that had already been sliced by three major highways and flooded over a significant part by the hydroelectric dam on the Tocantins River before I got there. The remainder of their habitat is now mapped for future development that will include mining, agriculture, logging, settlements, and more hydroelectric dams.[13]

The primates I found in Africa, Madagascar, India, and southern Asia were also sliding down the same steep and slippery slope in startlingly similar situations. Massive deforestation, for one reason or another. Hunting, to one degree or another. The douc langurs of Vietnam had the additional bad fortune to live in a place where the skies rained Agent Orange, a highly toxic defoliant.[14] Good-bye, douc langurs. The lion-tailed macaques of southern India, dedicated tree-dwellers dressed all in black save for a white lion's mane and tail tip, were reduced to somewhere between nine hundred and two thousand individuals as a consequence of tree cutting in the service of tea plantations.[15] Good-bye, lion-tailed macaques. The rarely seen gibbons of the Mentawai Islands, off the coast of Sumatra, might be thankful that many of the hunters on those islands are (or were recently) using a traditional hunting technology: bow and poison-tipped arrows. But shotguns are coming to the Mentawai Islands, so good-bye, Mentawai Islands gibbons.[16]

The details are different, yet the pattern remains monotonously the same. You and I both know, at least generally, why these animals are declining and going extinct. We also both generally understand how pervasive this trend is. Half of the world's few hundred primate species are threatened or endangered with extinction;[17] and the problem, we know, is not at all limited to primates. With the obvious exception of domesticated animals, every major group of large mammals on the planet is declining in numbers. Both you and I know, in addition, that in virtually every case the agent behind this powerful global trend of extinctions is the same: us. Why are animals going extinct? Since we already know the answer, and since we already know who is doing it, the question then becomes a little different: Why are people causing these extinctions?

The most immediate answer is that humans are competing with animals over resources. Sometimes the resource is the forest the animal lives in or the land the animal lives on. Sometimes the resource is a part of the animal—the ivory of elephants comes to mind—or even the whole animal. But in almost all cases, we can say that extinctions and extinction trends occur as a result of a resource competition between humans and animals. We are thus greeted with the current wave of extinctions and the current trend of species decline everywhere not because of anything new in the fundamental relationship between humans and animals, but because an ancient competition over resources has suddenly tilted in favor of humans as a result of two recent and entirely pragmatic events: first, the explosive rise in human numbers since the start of the twentieth century; second, the arrival of several powerful new technologies—bulldozers, chainsaws, shotguns, semiautomatic rifles, and so on.

The concept of a perpetual human-animal competition over resources seems reasonable, and behind it is an interesting human orientation. Let me express the moral aspect of this orientation in the way it was expressed to me one day while I was looking for the largest lemurs of Madagascar, the indris. I had found a number of them living in a small bit of officially protected forest, the Andasibe Reserve, located a few hours by train east from the capital city of Antananarivo.

Lemurs are mostly nocturnal primates, and those species now preferring the daylight still have a nocturnal ancestry, meaning that one characteristic of lemur anatomy is the tapetum lucidum: a reflective structure in the eye, located directly behind the iris, that has the useful effect of amplifying light by recycling it, and the possibly useless by-product of making the eyes sometimes seem to glow in the dark. Hence the name *lemur*, in honor of the Roman wandering spirits of the dead, the *lemures*.

The indris I found looked onto the world through preternaturally bright, lemon-yellow eyes, and they were about as big as medium-size dogs. They were colored in a pattern that alternated black with creamy white, and their faces were marked not only by those lemony eyes but by black, teddy-bear-style snouts and black and rounded, teddy-bear-style ears. Their hands and feet were long, about six times longer than broad, with toes that were more like fingers, and legs that seemed long and muscular. As the indris clung to the narrow trunks or near vertical branches of some small trees, those legs folded up; and when these lemurs moved from one tree to another, the power of their legs became apparent. Rather than leaping, these animals uncoiled the springs of their legs and sprang. Or bounced. Or pogo-sticked. They moved from one tree to another like arboreal kangaroos, launching themselves into parabolic trajectories that still enabled them to advance horizontally for distances as great as fifteen feet—or so it seemed to me, trying to estimate the distance from my perspective on the ground.[18]

At Andasibe, I stayed in a small, old-fashioned hotel that was part of the train station and included a small restaurant. In the station's restaurant one day, I ate a meal next to a couple of French families, diplomats and their offspring, I guessed. I chatted briefly with one of the men in this group, who, after I said I had been traveling around the world looking for endangered primates, declared, "You must have seen lots of the human primates on your trip."

After I agreed I had, he moved in for the kill: "I think they are the most important to protect, no? This is a sanctuary, is it not?"

That was a conversation stopper. My first reaction, after overcoming the more general sensation that I had just been insulted, was to think how wrongheaded his implications were. Andasibe is a tiny forest, slightly larger than three square miles, that at the time was (and for all I know still is) just barely protected. The entire annual budget for all of Madagascar's three dozen national parks and reserves was around one thousand dollars, which means that the island's rare primates, found nowhere else in the universe, were like the British crown jewels being carried around in a paper bag.[19]

But couched within the wrongheaded implications, like an edible nut inside a bitter shell, was a well understood foundational moral rule that anyone, myself included, would be hard-pressed to challenge. The foundational rule goes something like this: *In any significant competition between the interests of humans and animals, humans have the moral right to win. This is so because it is so.*

That brief exchange of words in the Andasibe train-station restaurant did not turn into an argument. If it had, I would have been seriously handicapped by the presence of that rule in the hands, or on the tongue, of my opponent. Because for most of us it appears to be profoundly and self-evidently true—what sane person would allow people to starve in order to feed a stupid little animal? How *dare* anyone worry about mere animals when so many millions, even billions, of people in this world are suffering so greatly?—I would never have thought to challenge it. Rather, I would have challenged my opponent's apparent belief that the facts, in this case, justified invoking the rule. In other words, I would have insisted that lemurs of Andasibe and the people of Madagascar were not, on balance, stationed at opposite sides of any significant competition.

At the time, Madagascar was certainly spinning in a downward economic spiral. I saw many obviously desperate people during my stay in the capital city, and while I was out in the country and looking for lemurs, rioters in the city tore through parts of the downtown, smashing plate-glass windows and in general expressing the fury that can accompany desperation. People were in need. Sometimes they were hungry. It might be argued that the indris of Andasibe could represent food for the people of Madagascar. But how much food? Those few lemurs would have made about a week's worth of meat for a small number of people. That was the one way in which indris and people were on opposite sides of a resource competition. On the other hand, I could produce a long and detailed assessment of the ways in which indris and people were on the same side, were and still are benefiting the Malagasy people. For one thing, these rare, attractive, exotic animals bring tourists and a steady stream of tourist dollars. They attract international attention and international donor money. They provide some important long-term aesthetic and cultural benefits. Like many other animals, moreover, the indris are intrinsically part of the island's complex tropical-forest ecosystem, which offers many critical environmental services, such as climate and soil stabilization, while maintaining a rich cache of biodiversity wealth that includes a thousand foods and ten thousand potential medicines for people. . . .

You see how it goes. I can only hope to win this hypothetical argument in the Andasibe train-station restaurant if I can demonstrate that living indris are worth more, in terms of human interests, than dead indris would be. "Indris and people are on the same track," I might summarize, before moving on to my dessert, only pausing briefly to add, "Their continued existence happily coincides with ours."

In any significant competition between the interests of humans and animals, humans have the moral right to win. This foundational rule may seem to you completely obvious. For most people, it's simply the way things are, have been, and ought to be. It is so because it's so. Indeed, this rule is so basic that we seldom think about it or even recognize it as a rule. It is also, perhaps unlike the other moral rules we do commonly recognize—the many *Thou shalts* and *Thou shalt nots* of this world—almost never written down. Unlike the other human moral rules, it is a rule by default. If you look back at all the written-down rules, the several prohibitions and exhortations that most people will agree express a fair sampling of human morality, you will see that in virtually every case the rules are about how humans should and should not treat each other. Animals, on the rare occasions they are mentioned, ordinarily fit into this system merely as human property. Since moral rules offer the potential for regulating how humans treat each other, they offer the promise of protection. Moral rules and morality create a protective umbrella, giving us, for example, some level of protection from murder, theft, and many other sorts of personal violation. But since these rules usually hold true for humans only, the umbrella does not ordinarily extend beyond our own species. Hence, the invisible rule, the rule by default.

That is the human moral perspective toward animals, and it should surprise no one. Indeed, a number of people have already called attention to this, and they argue with great fervor that our moral perspective *should* be constructed differently. Thus comes the idea of giving "rights," that is, moral rights with legal implications, to nonhuman species. The animal rights argument is, I believe, mainly an exercise in generalization. If we can say that membership in our own species automatically gives us certain moral and often legal rights, why not generalize that principle to include other species? Why not expand the moral circle?

I believe that in some circumstances, moral generalizations of that sort might work. But for now, let me just say that my goal is to describe things as they are, not argue how they should be.

The moral perspective is only one of several that, combined, make up what I will describe as a fundamental human orientation toward nonhuman animals. Among the other perspectives, I include the linguistic, psychological, and aesthetic ones, all of which overlap and reinforce each other like complementary twists in a complicated sailor's knot.

In the previous chapter, I referred to our linguistic perspective on animals, which is marked by several small speech habits, such as using the same pronouns to describe animals as we use to describe things. That unconscious linguistic habit overlaps and reinforces our psychological inclination to think of and see animals as if they really are things, rather than, say, complex biological entities with a psychological presence. Aesthetically, we often see animals in ways that might be compared to the way we see Victorian wallpaper: as something ornate and decorative but now, alas, faded and peeling off the walls, bound to be replaced by something cleaner and simpler and more modern.

I have come to consider this orientation as a large-scale, species-wide sort of narcissism. You may be used to thinking of narcissism as a standard form of ordinary egotism: the belief most people hold that they are better looking, more generous, more honest, more unbiased and psychologically open—even better drivers—than average.[20] Or maybe you think of narcissism as a more serious thing: a kind of irrational arrogance or the manifestation of a hopelessly immature personality or even a psychological disease. But I am using the word differently here to suggest not the abnormal affliction of an individual but the ordinary condition of a species. I think of it as the standard orientation of normal people, and I believe our minds work this way for evolutionary reasons. Thus, I call this kind of orientation *Darwinian narcissism*.

Perhaps the experience with maps first stimulated my thoughts on the subject. I found it interesting that the big map, the political map, so confidently advertised itself as representing "The World"—when the worlds of the animals I had come to find were simply not represented. I also began to understand that the "world news" on radio or television or in the newspaper was, likewise, really the human political news. Yes, there is plenty of animal news out there, such as the ongoing Ebola epidemic wiping out gorillas across a large swath of Central Africa. I came to recognize that whenever I heard about "the world population," I was really hearing about the world human population. And I began to conclude that whenever well-meaning people—for example, Bill and Melinda Gates, with their megabillion-dollar charitable foundation—set out to save "the world," they're thinking of the human world.

You may counter, with some wit, that only humans read maps, listen to the news, and understand the value of money. But I'm talking here about only humans, about the way we humans are oriented to see the world as consisting of only ourselves. Nor do I believe that this is entirely a trivial problem. Imagining that we exist alone in the universe, supremely dissociated from all other forms of life, is an aspect of believing that we are magically disconnected from our environment. Do that and then, one day, you begin to wonder why the air is suddenly bad, the water poisonous, and the climate starting to change in ways you hadn't anticipated. And when you set out to save the human world in the style of Bill and Melinda Gates and, as an accidental by-product of your narrowly focused efforts, destroy pieces of the animal world, you will also have compromised the future pleasure, health, safety, and equilibrium of the very humans you believed you were going to save.

Still, it may be in its most trivial manifestations that Darwinian narcissism looks most like . . . well, narcissism. Look up, next time you pass through a supermarket checkout stand. Look around, next time you walk into any magazine shop anywhere. The face you see reflected a thousand times over is yours, with some minor variations: again and again and again the face of your own species.

Here, I think, is how the nineteeth-century American author Herman Melville expresses the concept. The central problem in his novel *Moby-Dick* is not to hunt and kill a whale but rather to understand one. The book's crucial dramatic structure has less to do with action and more to do with language and perception, which may explain why no good movie has ever been made of it. It is too richly ambiguous for the cinematic medium.

That rich ambiguity results in part from the author's habit of draping the action beneath contrasting veils of perspective: those of several characters, including the captain, Ahab, and the first mate, Starbuck, as well as that of the narrator, Ishmael, who is himself a character with his own flawed vision. Ahab searches for an animal who has transgressed the human moral code. Starbuck seeks an animal who is a swimming piece of human wealth. Ishmael, an intelligent and impressionable young man, looks for an animal who represents a piece of human history, an object in human art, a curio of human science, and even, possibly, a mysterious deity bursting the mystical confines of some exotic human religion. The men on the whaling ship find whales, yes, but instead of seeing real whales as they really are, these characters see wavering phantoms rendered ungraspable by their own mental filterings and distortions, by the confusing reflection of their own selves.

A year or two after I finished my circumnavigation in search of the world's dozen most endangered primates, I met the well-known chimpanzee expert Jane Goodall. She was planning to write a book about the relationship between chimpanzees and humans. She had plenty of expertise on the subject but no time. I had no expertise but plenty of time. So we decided to combine forces to produce the book she had in mind. My part of the project began with some basic research, which included two extended trips to Africa to visit three major chimpanzee research sites and a number of other places where it was possible to find chimps. Back in the United States, I spent additional time finding caged chimpanzees, including those serving as biomedical research subjects in laboratories and others serving private owners as pets, substitute children, public entertainers, and photographic subjects for hire. During those travels in Africa and the States, I began to realize that the principle of Darwinian narcissism holds true for chimps as well.

It may be harder to see or demonstrate for them, since they don't have language to create their own self-referential labyrinths of myth and culture. They don't invent a thousand stories about themselves, decorate their world with a hundred million photographs of themselves, print magazines or publish newspapers about the drama of their lives, talk endlessly about what is going on in their world, or produce comprehensive maps of it. So my evidence for this conclusion about chimpanzees is simpler and less complete than the artifactual evidence for Darwinian narcissism that really surrounds *Homo sapiens*.

Yet it is still possible to point out, for chimpanzees, some examples of this interesting orientation phenomenon. These apes, now officially endangered but surviving in small pockets across the broad middle of the African continent, will respond in three basic ways to the sight of people. Since they are hunted for their meat in many places, the first and most common response is fear and flight. A second kind of response may occur in a few exceptionally remote places where people are a novelty. In this situation, instead of recognizing humans as dangerous predators, the chimps appear to regard them as audacious intruders. If the intruders are few and the resident chimps many, the latter may surround the former, screaming ferociously and, seemingly, working up the nerve to attack. For the human who finds himself or herself as the intruder surrounded by enraged and aggressive chimps, this can be unnerving—as I one day discovered. Fortunately, the chimps are also anxious. After all, they have just found in their forest a kind of creature they may never have seen before: a creature too big to be the prey but just possibly big enough to be a predator after all. The result is usually a standoff, with the screaming chimps gradually losing their nerve and fading away.

The third situation is about as rare as the second. It only occurs in about a dozen different places in Africa, the scientific research sites. At these places, good scientific study of the behavior of chimpanzees became possible only after the animals were habituated to the presence of people. *Habituated* means that the animals have concluded, after a long and difficult learning process, that the people they see before them are neither predators nor prey. You know that the chimps are finally habituated to people because they have by and large stopped reacting to them at all. Yes, they still see that people are there, and they still—as always—act as if they recognize some individual people and can certainly tell the difference between men and women. But for the most part, people, having proven over a long period that they are neither predator nor prey, have become invisible.

I mean *invisible* emotionally and psychologically, not literally, of course, but the ultimate result seems about the same. Habituation is an extraordinary event. Now you can follow a group of chimpanzees closely during their ordinary daily routine, starting from the moment they wake up in the morning, stretch, urinate off the edges of their sleeping nests, call out to each other, and climb down to begin another day. They see you but then they stop seeing you—don't notice, hardly care—and in this way, following them as that invisible presence, as the strange ape who does nothing, bothers nobody, and might as well not be there, you observe the unfolding of their ordinary lives. Habituation is possible only because their ordinary daily lives have almost nothing to do with you and nearly everything to do with them. They react to each other, follow each other, break away from one another, get in fights with one another, lie out in the sun and take a little siesta together during the hottest part of the day, and sometimes just take off in different directions with different ideas about where the best food or sex can be found.

So habituation has been accomplished the moment you stop being a distressing alien who might do something bad. Then you become not a friend but a ghost. As a ghost, a boringly insignificant presence, you are able to see for the first time the ordinary life of a wild chimpanzee, which is a vast and seemingly eternal drama involving other chimpanzees. That's Darwinian narcissism: the key to everything that follows.

CHAPTER 3

Definitions

Next: how shall we define the whale, by his obvious externals,
so as conspicuously to label him for all time to come? To be short,
then, *a whale is a spouting fish with a horizontal tail.*

—Herman Melville, *Moby-Dick*[1]

The documentary film *March of the Penguins* features the amazing lives of emperor penguins, who once a year emerge from the freezing ocean, queue up on the Antarctic ice shelf, and march seventy miles into the depths of an Antarctic winter in order to mate eagerly, gently, and (so it would appear) affectionately. They pair off with their chosen mates, then take turns protecting their fragile, fertilized eggs against the worst weather on this planet.

Emperor penguins, those fluffy little fellows built like bowling pins and strutting like movie stars in tuxedos, evoke an ideal morality we humans would like to see in ourselves: with romantic courtships, splendid monogamy, transcendent devotion to the fate of their offspring, and equal-opportunity parenting. It is an exciting vision, and for a moment we can see the possibility of animal values. For a moment, instead of asking why dogs and cats can live among people all their lives and still never acquire human norms, we start to wonder whether we should fly to Antarctica and try to acquire penguin norms.

Occasionally, then, the behaviors and values of another species resemble our own enough that we experience a flash of recognition . . . but mostly, the idea of animals possessing morality induces open incredulity or just plain bafflement. Open incredulity is the sign of a closed mind. Plain bafflement may be more a consequence of missing information; and the truth is that we are missing lots of information about the reality of most animals. We have only begun to study animal behavior, and we are limited because much of an animal's experience is physically remote and psychologically alien from ours. These are problems, limitations, barriers to seeing—but much more significant are the barriers of bias.

The linguistic bias I mentioned in chapter 1 is part of a larger intellectual bias that regards humans as fundamentally exempt from the systems and limitations of the rest of nature. The orientation bias I introduced in chapter 2 represents a way of experiencing the world so fundamental to the nature of who we are and how our minds operate that we are almost never aware of it. I called this orientation bias Darwinian narcissism, and I described it as the powerful, evolved tendency of each species to locate itself at the center of the universe. All animals, whether humans or dogs or elephants or sperm whales, exist in a self-contained world in which all significant actors are members of the same species. Whatever values they live by, therefore, are relevant primarily to themselves and are enclosed in a logically self-referential bubble. By our very nature, we are never fully capable of accepting an alternative value system as entirely valid—and so, in looking for animal morality, we make the first mistake, which is to search for a system of values that perfectly reflects our own.

This intellectual error, the search for human values made manifest in the lives of animals, is easy to parody. It's like dressing elephants in tutus. But in the wrong hands it can also become a serious or even a dangerous error. Theorists of a Western liberal inclination sometimes look for animals whose lives seem to manifest their own human ideal of an unusual generosity, sharing, cooperation, egalitarianism, and all-round niceness; and when they don't find such an animal, occasionally they try to create one—by, for example, promoting a vision of chimpanzees as nicer than they really are. Meanwhile, theorists of a harsher and more authoritarian bent sometimes look for creatures who seem to manifest high levels of obedience and self-sacrifice: the intense subordination of the individual to the welfare of the larger community. Honeybees might be a good example here, but do they represent the ideal model for considering human morality? The Nazis seem to have thought so, and thus they tried to enlist Europe's foremost expert on honeybees and honeybee communication, Karl von Frisch, to help promote their ideology.[2] Von Frisch would have none of it, and neither should we.

The medieval philosopher and Catholic saint Thomas Aquinas argued that "natural" was equivalent to "good," which meant that one could make moral judgments by appealing to naturally occurring examples. He based this notion on the premise that God, who created all of nature, could make only a nature that was morally good; and we see Aquinas's position echoed in contemporary times when people say that homosexuality is immoral because it violates nature.

The British philosopher G. E. Moore presented in his *Principia Ethica* (1903) a contrary position. He argued that ethical qualities such as "good" or "bad" are unrelated to natural qualities, such as "pleasant" or "desirable" or "attractive" or "contributes to my own survival." Moore solidified his stance by branding the opposing point of view as a formal logical fallacy: what he called the *naturalistic fallacy*. Moore's argument narrowly focused on the semantics of ethical philosophy, but the concept of a naturalistic fallacy is often more broadly interpreted to mean that one cannot find moral truths by looking at the productions of nature.[3] You can't say something is moral because it's natural, so this theory goes, and you can't call something immoral because it's unnatural.

Who is right, Aquinas or Moore?

Actually, neither. On the one hand, human and nonhuman moral systems derive from the natural process of Darwinian evolution. In that sense, then, our human morality is based on nature, and it can be explained and tested by reference to naturally occurring events. On the other hand, while evolution gives us and other social species morality in general, the particulars will vary according to any particular species' ecological circumstances and a complexly evolved response to them. So, while I will promote the idea of morality as a gift of nature working through evolution, let us not think of the lives of emperor penguins or chimpanzees or honeybees or any other animals as if they will manifest specific, particularized instances of what is right or wrong in the lives of humans.[4]

This creates a dilemma, however. If we can't expect to find perfect examples of human values manifested in the lives of animals, how can we possibly hope to talk about animal morality at all? What are we looking for? What is morality if not a system of values? And what other values are possible other than the human ones? Darwinian narcissism limits how far we are normally inclined to look for animal morality, and it often leads us to short-circuit our discussion through an approach I'll call *argument by definition*. Argument by definition means creating a definition that automatically proves your case. On the question of whether animals have morality, for example, this sort of approach would proceed by defining morality in ways that automatically limit it to humans:

- Morality appears only when people discuss or debate it.
- Morality can exist only if there are written rules.
- Morality is a symbolic system of values.
- Morality is what makes bad people good.
- Morality requires a religion to impose and interpret it.
- Morality is God's sacred gift to humanity.
- Morality happens only when you have high intelligence or rationality to grasp it.
- Morality is an aspect of human culture.
- Morality is a product of human civilization.
- Morality rests on "collectively shared norms," and such normative sharing is only possible with human language.[5]
- Morality requires an ability to feel guilt and a capacity for disgust, which are "obviously" limited to humans.
- Morality represents a solution to the Good Samaritan conundrum: the interesting problem of people living in large societies and interacting altruistically with others who are not genetic relatives.

Each of the above statements might be true or partially true, although several of them are debatable. Is it really true, for example, that only humans have a sense of guilt? Many animals show themselves capable of remembering very well what others approve and disapprove of. You might dismiss such animal memory as not even remotely associated with our refined sense of guilt, as, instead, merely a memory associated with the fear of future unpleasantness—but who's to say the human sense of guilt is not also associated with a fear of future unpleasantness?[6] You might say that only people feel the sensation of disgust, based on the observation that dogs love to roll around in the most disgusting substances. But have you ever tried to feed a dog something he or she doesn't recognize as valid food? Still, whether or not these several assessments and assertions are true or false, they all have in common the quality of inappropriately short-circuiting any discussion about the nature of morality. In each case, without inviting further discussion or additional thought, and even though we may find good equivalents in the psychology of animals, morality is presented as a humans-only entity by definition.

Darwinian narcissism makes it hard to define morality in any other way; and to be sure, we are members of a species intimately familiar with its own moral landscape. But asking a person to define morality is a little like asking an elephant to define a nose. "A nose?" the elephant begins. "Well, isn't it obvious? A nose is something at the end of your face that is magnificently long, perfectly round, miraculously flexible, very sensitive, and yet exceptionally powerful. It is long enough to reach your toes. It's rounder than any other organ. Its flexibility is a wonder to behold; and at its very tip, a nose has the astonishingly sensitive ability to grasp very small things. At the same time, a nose is capable of wrapping itself around an enemy and dashing him mightily to the ground."

One protests to the elephant, "But you have only described things that an elephant's nose has or can do. How about something like smell? Doesn't a nose also provide some olfactory ability?"

"Oh, yes," says the elephant. "Sorry! I forgot that part. Aside from its overall magnificence, its great length, roundness, flexibility, fine-tuned sensitivity, and powerfully destructive capacities, a nose will allow you to smell water from a dozen miles away."

One protests again, "But don't you get it? I have a nose, too. You must somehow change your definition so that it includes my nose as well as yours."

The elephant pauses, as if thinking deeply. "What? You mean that pathetic protrusion of flesh beneath your eyes and above your mouth? You must be kidding! Well, all right. I can accept that your tiny thing might represent some kind of earlier experiment that happened during the advance of evolution long before it reached its highest expression and logical conclusion with *Elephas maximus*. If you insist, I suppose we can call that thing a *pre-nose* or an *almost-nose* or a *proto-nose*. But surely we can safely say that such a minuscule excrescence is *not a true nose*."

You see the problem. This may at first appear to be a case of simple phylogenetic chauvinism, but the problem is much more fundamental to the experience of being an elephant. It is not merely the result of feeling special, full of pride for one's own species. Rather, it is the result of feeling unique and central to the universe. It would be easy to define *nose* in a way that works only for elephants, and any normal elephant will have great difficulty imagining any other definition. But such a definition, unfortunately, inhibits anyone from thinking intelligently about noses in general.

To think about noses in general, we need to drop the manifestational aspects of the definition: to stop being distracted by how this particular organ looks or is manifested in our own particular species. We need instead to try a simple functional definition. Every important organ has a core function that transcends the complexly different manifestations found in different species. So what is the simplest way to describe the core function of a nose? *A nose enables the sense of smell.*

Thinking about morality presents quite the same dilemma. We humans may have a moral organ that is equivalent to the elephant's nose: enormous, powerful, multifaceted. It is extremely easy to define morality by identifying this or that manifestation of human morality that might indeed be uniquely ours—written codes, cultural elements, intellectual analysis, an elaborate conscience, a fine-tuned sense of guilt—and thereby fail to recognize morality as it appears elsewhere, in other species. If it were possible to imagine morality or a moral system as an organ, like a nose, and if we could limit ourselves to considering its core function and thus produce a simple functional definition, what would it be? Try this: *The function of morality, or the moral organ, is to negotiate the inherent conflict between self and others.*

It sounds interesting, you might say—but what inherent conflict?

Let's think about that question based our own ordinary experience. You and I can go through days, even weeks perhaps, without experiencing anything that feels like a conflict with our fellow men and women. But since we all live in social groups of some sort, such a conflict will eventually, inevitably occur—even though, usually, that conflict will be anticipated and mediated by already existing laws or moral rules. (We might think of laws as formalized and officially enforced moral rules: for example, the legal prohibitions against murder and theft, which formalize moral rules known everywhere and encompassed in the Christian tradition of the Ten Commandments. In theocratic societies, moral rules and laws may be combined—or, rather, laws will be directly fabricated from the material of moral rules. In secular Western societies, the historical divergence of more formalized law from less formalized moral rule goes back several hundred years, but even after that extended historical split we can discern the common root for these two sorts of behavioral codes. In secular societies, laws usually cover the most obvious and easily defined possible violations, with official prohibitions against murder, theft, rape, and so on. Moral codes cover a fuller spectrum of possibilities, ranging from murder to deceit to disobedience, and they offer both positive exhortation and negative prohibition.)

The potential for conflict is always there, and virtually every single law or moral rule deals directly with the problems and potential problems of self versus others. You can choose to be honest—signaling direct information or intent to other people. You can choose to be kind—an act that requires others to be kind too. You can be charitable—to others who need charity. Or, looking now on the negative side of a moral potential, you can be cruel—but only to others. You can violate sexual taboos—other people's taboos. You can steal—but only other people's property. Are there moral behaviors, positive or negative, that a person can carry out in social isolation? Surely there are. One thinks of the mad bomber laboring alone in his dreary basement. But acts done in isolation are moral or immoral (positive or negative aspects of a moral system) only because of their perceived social consequence. They are perceived as ultimately resulting in damage to family, friends, or society at large.

So let's grant that living in any society inevitably places one's own self in conflict with others. Yet surely, you say, not all those conflicts are moral situations. What about the many trivial cases? What about, for example, the ordinary behavior of walking along a sidewalk in a town or city? Walking along a sidewalk is not a social act, but let's assume that someone else is walking toward you on the same sidewalk. Ordinarily, sidewalks are built wide enough for two people to pass in opposite directions with little physical negotiation, but what if the other person has moved to the wrong side, thus momentarily slowing down your progress and otherwise distracting or even inconveniencing you? You and the other person are now coming into a situation of potential conflict. Does that situation induce a moral response?

It depends on the seriousness of the full event. Let's first imagine the event is not serious. The other person blocked your way accidentally and signals that she did so with a quick apology: "Oops. Sorry." This conflict has been negotiated through manners. The other person displayed good manners. Or, if she failed to apologize, we might call it bad manners. Alternatively, however, we can imagine a somewhat similar situation that turns serious. The other person deliberately blocked your way, glared at you, and when you attempted to proceed, he actually shoved you aside. This conflict has obviously become more serious than the first, and we might believe the other person has done something that moves beyond a problem of manners into the arena of morality. Perhaps we can say he has shown negative moral behavior. He's done something reprehensible, something bad. In the meantime, though, you recover and thoughtfully conclude that the other person somehow deserves not an enraged response but magnanimous forgiveness. You have shown what many people will regard as positive moral behavior.

Manners and morals inhabit the same general territory. They are both ways of negotiating the conflict between self and others, and they gradually merge into each other at the place where the trivial meets the serious. You might say that "trivial" and "serious" are in the mind of the beholder, and indeed we can endlessly debate precisely where manners end and morals begin. This is one of the areas where people from different societies and cultures often fail to agree. Is a scantily clad person walking down the street presenting a problem of manners, morals, or neither one (no conflict whatsoever)? It depends on whether you live in Boston, Baghdad, or Paris.

Still, we all know intuitively when we regard something as an issue in morals rather than manners. When we experience a moral opinion, we feel a characteristic kind of emotional urgency. It is as if, to recall an analogy suggested by cognitive scientist Steven Pinker, a toggle switch has been flipped to the *on* position, and we feel ourselves in a moralizing frame of mind, a mental state of *moralization*. "Moralization," Pinker writes, "is a psychological state that can be turned on and off like a switch, and when it is on, a distinctive mind-set commandeers our thinking."[7] That mind-set, having distinguished *wrong* from *awkward* or *ill-mannered*, includes the urge to punish the person or persons perceived as being *wrong*. This same moralization state, of course, having similarly distinguished *right* from merely *nice*, might include the urge to praise or encourage or otherwise reward people who have acted rightly or righteously. So the urge to punish is not in itself a defining element of moralization.

By comparing the thinking of vegetarians who refuse to eat meat for ethical objections to animal suffering with that of vegetarians who abstain for practical reasons related to health, psychologist Paul Rozin has attempted to envision the larger dimensions of the moralization mind-set. Ethical vegetarians, Rozin found, more often considered meat a noxious or contaminating substance, so they were unwilling to eat food with even a trace of meat or a tiny drop of meat juice. They were also more inclined to promote vegetarianism in others, and they were likely to conflate their vegetarianism with other perceived moral virtues, such as peacefulness.

The idea of a moralization switch, a toggle that will turn on a particular set of emotions in the brain, conforms to what we understand about a few culturally induced changes in our moral codes. Getting a divorce and having illegitimate children were both, until recently in our culture, likely to induce the moralization response, yet today these actions are generally regarded as not within the moral sphere. We might say they have been *amoralized*. In the meantime, activities such as smoking have recently provoked the switch to flip in the other direction, taking us from an amoralized to a moralized state. As psychologist Rozin has noted, people in the recent past avoided smoking because they didn't enjoy the experience or they suspected it could harm their own health. But once the idea gained currency that secondhand smoke was a threat to the health and lives of others, the toggle was flipped to the *on* state. Smoking suddenly became an act likely to inspire censorious behavior in others, so that smokers are now avoided or, at least, required to conduct their filthy habit at a remove. Tobacco smoke itself is understood as a contaminant, enough so that hotels now dedicate entire floors to the nonsmoking modality. And tobacco companies have become liable to punishing lawsuits. So in our society, cigarette smoking has gone from a rather trivial issue of manners to a somewhat serious one of morals.[8]

We should not be diverted by these few examples into thinking that morality is a mere cultural invention, a series of random and arbitrary conventions that have mysteriously appeared for random historical or social or cultural reasons. No, clearly, morality is a universal experience among humans. We will not find a society on Earth where people fail to experience similar emotions in similar situations: moralizing emotions in response to murder, or to theft or incest or such profound personal violations as rape and violent assault. Murder, theft, incest, rape, assault: These are all unquestionably serious events, universally understood to evoke the moralization response. In comparison, such behaviors as smoking, in Rozin's example, may be either trivial or serious, depending on how much we believe secondhand smoke represents an assault on others. Because the point where the trivial becomes serious is not always clear and can occasionally shift according to cultural developments and new understandings, trying to identify that point precisely would be a waste of our energy. Let's just say that someone's behavior is *serious*, in this context, when it is understood to affect significantly the welfare or potential welfare of someone else.

Given this extra observation, I will modify the functional definition of morality in the following way: *The function of morality, or the moral organ, is to negotiate the inherent serious conflict between self and others.*

So far I have been using the concept of a moral organ as an analogy, but let me now suggest that it could be something more than a simple analogy. Ordinarily when we think of organs, we think of a visible and tangible thing: an eye, nose, or ear we can touch, a liver that can be damaged by disease and replaced by a surgeon. But we also recognize that most organs actually include significant parts that can't easily be recognized or touched or even seen. Our organs for sight, for example, consist of not only the accessible outer orb, with its many easily distinguishable parts, but also a less accessible and less visible cable of neurons that quickly expands and merges into an even less accessible, virtually invisible, and very complex electrochemical system that ultimately enables our brains, and our minds, to translate neural information into the coherent and meaningful miracle of sight. Some of the visual organ is visible and tangible. The most complex and arguably most important part is not. The moral organ I am imagining would consist of material that resembles the least accessible, least visible portion of the visual organ. That is to say, the moral organ is essentially neurological in nature.

We tend to believe in the uniqueness of our own human organs just as much as we believe in the uniqueness of our species, but the vast majority of such organs appear in similar form among many other species. Eyes? This organ first appeared in rudimentary form during the Cambrian epoch around 550 million years ago and may have represented such an advantage, according to the thinking of zoologist Andrew Parker, that it caused an evolutionary explosion during which the zoological kingdom expanded in diversity from three major groups, or *phyla*, to thirty-eight.[9] Noses? What mammals don't have them? Ears? Livers? An evolutionary understanding enables us to appreciate the trans-specific history of organs. Most of them emerged a long time ago, and they have proved to be so useful that they have been retained by a large genetic tree that has over deep time branched and spread tremendously. This truth gives us the best answer to the old chicken-and-egg riddle. Which came first: chicken or egg? We know from many well-preserved fossils that dinosaurs laid eggs and hatched out of them far earlier than sixty-five million years ago, the date marking the end of the big dinosaurs. We also know that chickens appeared only about ten thousand years ago, when people living in what is today Vietnam domesticated a hybrid of two wild birds, the red and gray jungle fowl. So eggs came first, because they're trans-specific.

The moral organ I am imagining may reside primarily in the limbic system of the brain, that large association of neurons located within the deeper layers of the cerebral cortex, although clearly it interacts with many areas of the brain and in any case need not have a specific form or limited location at all. Neuropsychologist Paul MacLean memorably promoted the idea of a limbic system in his book *The Triune Brain in Evolution* (1990).[10] MacLean argued not merely that the neurocircuitry of the limbic system had been established through evolutionary processes, but also that the fundamental structure of this system evolved long before the evolution of the distinctively human neocortex. Because of its ancient evolutionary roots, he declared, a similar limbic system will be found in the brains of all other mammals. The limbic system generates the affective part of our experience (feelings of sadness, joy, love, hate, anger, excitement, and so on), and it gives emotional significance to information flowing into the brain from the senses (sight, hearing, smell, taste, and touch). Affects—what we more commonly call *emotions* or *feelings*—are the internal value-coded experiences that enable us to identify the reproductive and survival value of the things we might wish for or hope to avoid, and of the actions we might take or refrain from taking.

It would be easiest if one could with complete assurance simply name all the moral emotions, then identify the biochemistry and neurocircuitry identified with each. Certainly we can name some of them, at least for human morality. They would include empathy, guilt, shame, pride, righteousness, a sense of justice, an urge to fairness, an inclination for revenge, and so on. But that list is apparently incomplete and surely simplistic, just as our knowledge of the biochemistry and neurocircuitry of the emotional brain is in its infancy. Moreover, the moral emotions are all situational. That is, they're triggered by particular situations. They're also divergent according to our relationship to the moral act in question. One set of emotions becomes appropriate when we're the actors. We might feel guilt or shame at some wrong thing we've done, or pride and righteousness if we've done something right. But when we're watching someone else behave in moral ways, we might feel an entirely different set of emotions. If the other person has done something we consider unfair, for example, we might experience a desire for justice or an urge to fairness. So the emotions governing or promoting moral behavior, the emotions that make up what I am calling the moral organ, are complicated indeed.

An additional issue is whether any of the moral emotions are trans-specific. A common, traditional belief has been that emotions in general (whether moral or otherwise) are not trans-specific—that only people have emotions of any sort, whereas animals don't. But of course animals do have emotions. All the best experts on elephants, including the real pioneers in field research on elephants—Iain Douglas-Hamilton, Cynthia Moss, and Joyce Poole—have described many instances of emotional behavior among elephants, and all three have experienced situations that convinced them that elephants feel, for instance, grief.[11] The best experts on other animals will likewise readily describe emotions they believe they have seen expressed. Since emotions are hard to measure, our strongest evidence at the moment for animal emotions may be the observations and opinions of field biologists. Any properly skeptical scientist, especially one who has not spent time with animals in the field, may object, *We don't really know*. I respect the skepticism. I will also note, however, that the skeptical scientist locks up his or her lab at night, goes home to the beloved pooch, and starts talking baby talk. Our favorite dogs and cats have emotions close enough to our own to make us respond with abandon and utter confidence in the authenticity of the experience; and if they have emotional problems, the veterinarian is likely to give them the same psychoactive drug—Prozac, for example—she takes herself.[12]

My general definition of morality, then, includes the possibility not so much that animals—at least mammals—have moral systems analogous to our own (that is, have neurologically based structures located in the limbic system and consisting of a complex mixture of situationally sensitized emotions comparable to what we possess), but that they may have moral systems homologous to ours. *Analogous* describes a similarity that could be coincidental. *Homologous* describes a similarity derived from common origin.

The difference is significant. Take the case of rat laughter, for example. Or should I say rat "laughter"? Neuroscientists Jaak Panksepp and Jeff Burgdorf began to consider the matter after they had spent some time watching their laboratory rats play. That the rats were actually playing was so obvious neither researcher thought to put the word in quote marks. It was just *play*, the sort of thing we're all familiar with because we've seen plenty of animals, domestic and wild, do it. But, Panksepp and Burgdorf wondered, was there an auditory element to the rats' play that they were missing? They asked the question experimentally by deafening some rats and then observing their play, thereby discovering that being deaf moderately reduced the intensity of play. This implied not only that the play had an auditory component, but that such a component encouraged or promoted playfulness.

Since the auditory event was apparently happening outside the range of human hearing, the researchers next brought in some special equipment designed to register ultrasound: vibratory productions too high in frequency to be detected by the human ear. Eavesdropping thus, they discovered that the rats' play was accompanied by an abundant production of *chirping* made in the range of fifty kilohertz. These chirpy vocalizations, it turned out, were particularly abundant just before play began, as if they were associated with anticipation or encouragement, or as if they might be considered "a general index of affectively positive incentive motivation."[13]

By this point, Panksepp and Burgdorf had decided that the high-pitched chirping might best be described as a "laughter-type response," and they began to pursue that concept further. So in the spring of 1997, Panksepp arrived at the laboratory one day and said to his junior partner, Burgdorf, "Let's go tickle some rats." They tickled rats and quickly found that the fifty-kilohertz chirps were being emitted at more than double the abundance ordinarily recorded during play. They also discovered that the animals seemed to seek out being tickled, and that this activity stimulated them in their own play. Thinking in evolutionary terms, the scientists decided, as they would later phrase it, "to remain open to the possibility that there was some type of ancestral relationship between this response, and the primitive laughter that most members of the human species exhibit in rudimentary form by the time they are three months old." The primitive laughter found among babies and young children develops over time and is "exquisitely expressed" when children reach the age where they begin their earliest sorts of social play, particularly in chase games such as tag.[14]

This was a fascinating observation, but a number of Panksepp and Burgdorf's scientific peers didn't think so, and many insisted that one can never know if animals have feelings (or "affectively positive incentive motivation") at all. The observations on rat laughter, or "laughter," were first reported by the popular media: featured in a BBC program and on the Discovery channel. Following that popular exposure, Panksepp received many letters from rat enthusiasts, including one from a young woman from California who, after seeing the Discovery show, tried tickling her young son's pet rat, Pinky. In only a week Pinky was conditioned to expect tickling, and the woman would occasionally be able to hear the high-pitched chirping. "It's been about 4 weeks," she wrote, "that I have been tickling him every day and now, the second I walk into the room, he starts gnawing on the bars of his cage and bouncing around like a kangaroo until I tickle him." Her rat preferred being tickled to eating, she went on, and she noted that he would turn over and expose an especially ticklish spot on his stomach. "It's the funniest thing I've ever seen, even though my family thought I had lost my mind until I showed them."[15]

Meanwhile, Panksepp and Burgdorf were encountering a similar response from some colleagues about their own rat-tickling proclivities. Indeed, they had already experienced both the scientifically appropriate skepticism and a more automatic, unthinking fear of anthropomorphism after they reported some earlier studies on the distress vocalizations produced by baby chickens, guinea pigs, and puppies after being separated from their mothers. The distress vocalizations, Panksepp and Burdorf had speculated then, might be considered animal homologues to infantile crying in humans.

As for the rat experiments, the researchers went on to identify a number of ways in which fifty-kilohertz chirping was comparable to social laughter in young children. It was provoked by play and tickling. There were tickle-sensitive parts of the body. As the rats grew older, they were less susceptible to being tickled. There was a negative relationship between unpleasant experiences and the capacity to chirp, and a positive one between chirping and play. Tickling was more attractive to rats with higher levels of fifty-kilohertz chirping, while rats who exhibited more abundant chirping and a greater response to tickling were more popular—that is, more likely to be chosen as social partners—than other rats. And so on.[16]

Panksepp and Burgdorf's response to the reflexive fear of anthropomorphism on the matter of rat-tickling and the possibility of rat laughter might be summarized like this. First, no one had so far come up with a better assessment of the phenomena they were describing. Second, it seemed to make perfect evolutionary sense to believe that "affective processes" could be found in the brains of mammals other than humans. And third, if they were indeed witnessing an animal homologue to human laughter—that these fifty-kilohertz chirps were "the sounds of social joy"—then they had found in their laboratory rats an "excellent animal model to help decipher scientifically one of the great mysteries of human emotions—the primal joyful nature of laughter and positive social interchange." The chirping of rats that occurs during play or while being tickled in the ribs, then, might not be just an interesting analogue to human laughter; instead, it might be a homologue to it—and we can therefore think of it not as "laughter," but rather as laughter.

Inherent in my definition of morality, for both humans and other animals, is the idea of conflict, which requires individuality. Individuality comes in part from genetic diversity, and for some taxa—ants, for example—there is minimal conflict partly because there is a minimized genetic diversity. The socially active ants in a colony are sisters descended from the same mother, the queen, and most conflict among ants plays out between different colonies or different species, rather than between individuals of the same species within a colony.[17] This reduction in individuality among ants makes them remarkably effective as harvesters, predators, and warriors. If you were able to collect all the millions of individuals of a single African driver-ant colony, you would discover a predator weighing perhaps forty kilograms—equivalent to a medium-size predatory cat—but with the extraordinary ability to change shape, spread broadly, advance en masse at an approximate speed of twenty meters an hour, and yet smell, attack, and consume in millions of different places simultaneously.[18] That image suggests, according to peripatetic myrmecologist Mark Moffett, "a superhero" not yet invented, a crime-fighting boy "who can disassemble, such that his hands can stop a crime while his head commutes to the office to write up the news report—Superman and Clark Kent both at the same moment."[19] That's remarkable, but the superhero coheres in part because of a general absence of individuality among the organisms making up the greater whole.

Also inherent in my definition of morality, for human and nonhuman alike, is the idea of choice. Morality requires distinctive individuals capable of making distinctive choices, moral choices that favor the interests of either the self or others. That's the source of tension in morality, and the logic for any moral system's existence. We believe we understand our own process of moral choosing, and we imagine it includes various internally experienced mental events. But what about animals, whose minds are mostly inaccessible to us? When we look for choice-making among animals, we are looking for behaviors that seem to demonstrate flexibility.

Flexibility means the ability to challenge one's normal habits or to tease the usual, to choose from a variety of options—and for many mammals that flexibility may best be displayed in play.[20] Indeed, we might like to think of play as an echo or mirror image of moral behavior.[21] The rules for both moral and play behavior are largely imaginative constructions, to recall a phrase and concept favored by cognitive ethologist Marc Bekoff in his book *Wild Justice* (2009). And like moral behavior, "social play is a voluntary activity that requires that participants understand and abide by the rules. It rests on foundations of fairness, cooperation, and trust, and it can break down when individuals cheat."[22]

Play typically includes a special sign or indicator, as if the participants need to be reminded that they're playing, and sometimes it begins with an introduction, as if such is necessary before the curtains can part and the stage be revealed. As we have seen, rats chirp or laugh as a prelude to play.[23] Apes and other primates may also express their pleasure in play with laughter or chuckling (shallow and rapid breathing that can be silent or vocally expressed) and with a smilelike *play face*, which is a relaxed, openmouthed expression that in most species conceals the teeth but in a few others (gelada baboons, sooty mangabeys, lion-tailed macaques, Sulawesi macaques, humans) displays them.[24]

Canids—including wolves, coyotes, wild and domestic dogs—show a relaxed and smilelike play face, but they more overtly signal an invitation to play with an approach centered on what Bekoff describes as the *play bow*. Anyone who owns a dog should know what the invitation to play looks like. It begins with one dog approaching a potential partner in a relaxed and perhaps mildly exaggerated style. Then, while keeping the hindquarters raised, the play-intending animal lowers his or her head with a quick, crouching drop of the forelegs. The dog barks and tail-wags, both actions done in a relaxed manner that adds to the impression of friendly intent; but the bow may be the most emphatic part of the approach, while the approach and bow jointly serve as an invitation: *Let's play.*

The play bow, according to Bekoff, also serves as a before and after reminder, an apology of sorts, in case things get out of control. Play for dogs is a rapidly shifting series of behavioral sequences that, in nonplay circumstances, are part of a dog's serious repertoire of fighting, preying, fleeing, having sex, and so on. During play, dogs seem to be reminding each other that even though their actions may seem serious, they are not. A reminder could be especially important in biting accompanied by headshaking. A dog biting while shaking his or her head rapidly back and forth is ordinarily going for the kill in a real predation or trying to incapacitate an opponent in a serious fight for dominance. In play, however, a dog will introduce the biting and head-shaking sequence with a quick bow, as if to say, *This might hurt, but it's still play.* Then, following the biting and head-shaking, the dog will once more bow, as if to apologize or to finish with a reminder that it was all in fun. Playing canids inhibit the intensity of their bites, since a real bite can break the skin and do serious damage. At the same time, though, routinely signaling that it's play is a way of keeping the imaginative world of play from breaking into the real world of serious life, a way of making sure that happy rough-and-tumble doesn't unhappily turn into tough-and-rumble.

So animals know they're playing because they inform each other of the fact—with special vocalizations, facial expressions, postures, and styles of locomotion. Not very different from the various ways we humans remind each other that we're just playing. But how do we humans know when animals are playing? Often, *we just do.* The play of animals so resembles our own that we intuitively know what's going on. Intuition, the sudden grasp of a probable truth, leads us to describe this strange behavior of animals with the same word we use to describe our own strange behavior. The behavior is *strange* because play is a source of profound pleasure, which means that our brains are telling us how important it is, yet it looks on the surface to be marked by high expenditures of energy for no important purpose whatever.[25]

Larger-brained species, or those with a comparatively bigger neocortex, seem also to be the more playful ones,[26] but the most overtly playful species are still dispersed broadly among mammals, so that we can talk about playing apes or monkeys, rats and mice, kangaroos and wallabies, whales and dolphins, as well as playing dogs and cats and their wild relatives among the canids and felids.[27] Such broad dispersal implies a deep history, and it leads us to ask why playfulness should have evolved so long ago and been stably maintained by evolution for such a long time. Animals and people play, presumably, for the same immediate reason. Play is fun, a source of profound joy. But why has evolution made play fun—indeed, so much fun that we can hardly resist the urge or invitation? Play is costly, in terms of energy spent and time lost and an increased risk of being preyed upon. So what benefits justify the costs? Why should mammals spend anywhere between 1 and 10 percent of their time at play?[28]

Some of the benefits of play are physical. A playing animal exercises and thereby increases strength, endurance, agility, and alertness. The playing animal exercises the important physical and mental sequences involved in fighting, preying, fleeing, and mating—sequences that may be essential at some sudden and unexpected moment during the ordinary unfolding of real life. So play is useful for survival and well-being, but for group-living animals, play also helps to develop social skills.

The social aspects of play seem to reinforce some of the virtues associated with what I call *attachments morality.* Social play routinely provides an excellent model of reciprocal cooperation, for example. The play may provoke a steady rhythm of turn-taking—my turn, your turn, my turn—that can be quite distinctive. Smoke chases and play-bites Spike, and Spike returns the chase and play-bite, and so it goes. This is mock fighting, and the easiest way to distinguish it from real fighting is to note the turn-taking: a regular pendulum of role-playing that oscillates steadily between aggressor and aggressed-upon. Even when the playing individuals are radically different in their abilities—say, a person capable of throwing a ball and a dog skilled in fetching it—their joint play can quickly devolve into the distinctive rhythm of reciprocity.

Back-and-forth, your-turn-now-mine, is the sort of reciprocal cooperation that so marks social play, and we might imagine that it includes some inherently understood rules promoting the fairness required by reciprocity. Certainly, written fairness rules are an essential aspect of officially organized games for people; and anyone who has watched dogs in a play group at the park will recognize the animal who, naively or neurotically unattuned to the usual rules of the game, quickly inhibits play among the others. But reciprocal cooperation and the structure of fairness that sustains it are still both aspects of attachments morality, while social play also teases or tests and ultimately reinforces what I call *rules morality.*

Attachments morality is based on some general principles favoring cooperation and kindness (or altruism), while rules morality emerges as a series of more particular responses to social conflict. I will in future chapters consider at length the five realms where such social conflict is likely to be most intense and the responding rules most emphatic. Those same realms—of authority, violence, sex, possession, and communication—are still important during play, but their relevance is often relaxed or mitigated, sometimes reversed, and so the play rules become flexibly distorted and even comic reflections of the moral ones.

During play, for example, a dishonest communication may be incorporated into the game, and so we find that a fake ambush simply adds to the fun. Likewise, the rules of ownership may playfully be contested. One participant may tease the other with an interesting object, refusing to let go and thereby provoking a chase. Or two playing animals may contest ownership of an object in a tug-of-war. The rules of sex, who mates with whom and under what circumstances, can also become relaxed during play, and so you might witness an experimental combination or a mock sexual act, one individual briefly mounting the other in bright enthusiasm but absolutely no engagement. Just as children will pull out their toy guns and point them at each other for the sheer joy of running, chasing, and shooting someone else dead, all in the spirit of fun, so animals will playfully unleash predatory acts against one another. It's as if what's forbidden becomes enticing, and so we have, for some animals (including rodents and some primates), the game of attack-and-tickle, and for others (for example, domestic and wild cats) that of stalk-and-pounce.

In chapter 5, I will describe the rule supporting authority as both self-referential (the rule supporting authority is sustained by the authority) and distributive (the authority rule is essential in supporting all the other rules). That's a fancy way of saying that the authority rule is the first and most important rule; and when we examine the games animals play, we find an abundance of activities that playfully test or even temporarily reverse authority, although in doing so they finally reinforce it.

Testing or reversing authority cannot easily be done without the embodiment of that authority—the dominant individual—cooperating. How? By practicing self-restraint. *Self-handicapping* is the term introduced by Bekoff, and the ordinary restraint we often see in play—coyotes biting with less than maximum force, say—is a form of self-handicapping. But self-handicapping becomes more obvious when two playing individuals are of obviously different size or social rank, and the larger or higher-ranking one actually reverses roles with the smaller or lower-ranking individual.[29] You and I are familiar with this kind of role reversal because it happens frequently among humans—when, for example, a father arm-wrestles with his son or daughter. No matter who finally wins, the good father carefully and somewhat dramatically makes it seem like a real contest. It happens also among eastern gray kangaroos and gray dorcopsis wallabies, between mothers and their young.[30] It happens among gorillas. A huge silverback begins playfully wrestling or boxing with a young juvenile one-tenth his size. The juvenile reaches up to punch the old guy in the nose, and the giant male—towering and hunched over, showing his openmouthed play face—backs gingerly away from the mighty blow.[31] I've seen it as well among wild chimpanzees: an old male playing ring-around-the-rosy with a two- or three-year-old, both of them laughing and equally engaged when, in reality, the older one could simply have picked up and tossed the young one aside. Among wolves, a dominant male might roll over onto his back in the middle of a play fight with a subordinate, making himself openly vulnerable in a way that would never occur in a serious fight between the same two animals.[32]

The point of self-handicapping and role reversal is not, finally, charity or generosity or a spontaneously occurring impulse of kindess. It is not, somehow, part of evolution's secret plan to produce a better world, to promote human or animal societies that are nicer, less hierarchical, more egalitarian.[33] Morality is not niceness. Egalitarianism is not inevitably a moral goal. On the contrary, maintaining the established hierarchy is an important part of most moral systems, which include rules to reinforce authority and the hierarchy of social power distributed around it. Self-handicapping and role reversals simply make play between socially unequal partners possible, and if anything, they serve to remind everyone of what, in actual life, the real hierarchy is and where, once play ends, the real authority resides.

CHAPTER 4
Structures

Ahab: "All visible objects . . . are but as pasteboard masks.
But in each event—in the living act, the undoubted deed—there,
some unknown but still reasoning thing puts forth the
mouldings of its features from behind the unreasoning mask.
If man will strike, strike through the mask!"

—Herman Melville, *Moby-Dick*[1]

Kevin Weeks, ex-boxer, loan shark, extortionist, and thug, kept his day job as a bouncer at the Triple-O bar in South Boston. But Weeks was also a member of the inner circle of the Irish mob from the 1970s to the late 1990s and a hard-knuckled enforcer for the infamous James "Whitey" Bulger and Stephen "the Rifleman" Flemmi. During his criminal career, James Bulger murdered around forty people. Stephen Flemmi was involved in around thirty murders. Kevin Weeks participated in several of those crimes, but life in the mob was not always an exciting series of plots, fistfights, shakedowns, and killings. In his chilling confessional memoir, *Brutal* (2006), Weeks claims that much of the time his existence was every bit as boring as yours and mine. "On a typical day, Jimmy and I would spend an hour a day with Stevie, doing business. I mean, we weren't doing crimes every day. We weren't animals, and except for the business aspect of our lives, we had boring, regular lives."[2]

We weren't animals. Human murderers may be criminals of the worst kind, and when we want to emphasize their apparent lack of ordinary moral values or restraint, we sometimes call them "animals"—but that comparison even the basest professional criminal will reject. Criminals are people, too. Criminals have values, too. Ordinary criminals share the ordinary idea that morality is a remarkable thing possessed or potentially possessed by every single person on this planet simply by virtue of his or her membership in the human species.

Where does this impressive, humans-only thing come from? How did we manage to be so blessed? The more traditional idea about the origin of human morality regards it as a code or a system of positive behaviors imposed from the outside. This is what I call *morality's external narrative*, and it is typically based upon one of two firmly held theories: the sacred gift and the social contract.

We all recognize the theory of the *sacred gift*, which suggests that morality was handed down by an all-powerful deity. In some versions of the theory, the deity looked down upon an inherently sinful, badly behaving humanity, was profoundly dissatisfied by the sight, and put in writing a moral code for people to live by and thus change the tenor of their behavior. Although people who spend their lives as professional interpreters of the sacred gift—ministers, pastors, priests, imams, rabbis, gurus, and the like—are unable to agree with one another on its precise details, the general outlines are clear and amazingly consistent between religions and across cultures. Everyone knows that murder, theft, and deceit are bad, that submission to proper authority and following certain tenets of sexual propriety are good, that we should treat others as well as we would like to be treated, and so on. Of course, the cross-cultural consistency of the sacred gift might well suggest an ecumenical interpretation: that the deity who supplied morality is universally known and worshipped but given different names in different parts of the world by a confused humanity.

The second theory supporting an external narrative for morality's origin is a secular one: the idea that a positive moral code has historically been imposed upon neutral or naturally antisocial individual people by the community. This we can call the theory of a *social contract*, in deference to a concept promoted by European philosophers John Locke, Immanuel Kant, Thomas Hobbes, and Jean-Jacques Rousseau, and named in Rousseau's influential book *The Social Contract* (1762).[3] If we dress up this premodern theory in contemporary clothes, it might be called the theory of contractarianism or, more broadly, of cultural construction or cultural evolution.[4] Both the social contract theory and its more contemporary variants often imagine positive morality as a language-based code. They imply that the origin of positive morality was a historical event, albeit a gradually developing one. And they tend to avoid the question of universality. How do we explain the universality of human moral behavior?

These two theories might seem to be radically different: one sacred, the other secular. But they have in common the presupposition that humans emerge from the womb unendowed (or decidedly underendowed) with morality. Infants emerge into the buzzing chaos of the world either behaviorally negative (a condition of sinfulness as an inherited consequence of the Original Sin or some equivalent) or behaviorally neutral (as a result of being born with a mental Blank Slate). In either case, since human infants come to us without morality, they need to acquire it or, less actively, be placed in circumstances conducive to receiving it. So morality is an external thing that must somehow be imposed or received, and without that imposition or reception, so the thinking goes, we would all be left minus the mark of human civilization or the milk of human kindness. We would be left in our original state of darkness: as base, anarchic, self-involved, ignoble, bloody, and irrationally violent as mere animals. In short, both of these theories show us a vision of morality as an external code that only humans can hope to possess.

There is, however, another narrative for the origins of morality—an internal one. This internal narrative suggests a third and less traditional theory: that morality is something you already have because you're born with it, at least partially or roughly or rudimentarily. Earlier versions of this idea might have said that morality comes from the heart, not the head. In speaking of "the heart," of course, one is really referring to feelings or emotions. It's a psychological theory of morality, and the version I promote in this book is based on the idea that our psychological nature is significantly shaped by past events—not so much by events from the recent and brief past of childhood, but more events from the ancient and extended past of our evolutionary beginnings as social animals. This internal narrative offers a theory of morality as an evolutionary development.

It is a common if naive complaint that ascribing an evolutionary aspect to behavior implies the development of "hardwired" mechanisms. The term *hardwired*, recently borrowed from the vocabulary of computer science, produces a mental image of electronic circuitry inside a machine; and the routine use of the term can create a crude caricature of how evolution shapes behavior. It's important to recognize that evolution can have exquisitely subtle, intricately complex, and yet very powerful effects on behavior, producing not a few simple on-off events so much as an immense array of predilections and potentialities that will often still be fully responsive to reality, reason, and learning.

Evolution may not even directly translate into behavior. Rather it can translate into emotions that in turn promote behavior. Consider the common household gerbil. These little animals make good pets, but they like to dig, and in the normal caged environment this interest may turn into an obsession, so that they're spending up to a third of the day trying to dig a hole in the corner of the cage. This obsessive behavior is driven not by some abstract sense of the importance of digging. Rather, it's the consequence of the emotional need for an underground tunnel and nest: a safe home. Gerbils with such a feature added to their caged environment will stop digging, since the emotional need has been fulfilled.[5] "Animals don't have purely behavioral needs," concludes animal scientist Temple Grandin, in *Animals Make Us Human* (2009). They have emotional ones.

Nor will some behaviors appear at all without the appropriate environmental or learning event that promotes the necessary emotion. Wild rhesus monkeys ordinarily show a strong fear of snakes, and we can ask whether they've inherited that fear or learned it. But that may well be a misleading question, since it assumes the answer will be either one or the other, whereas it could be a little of both. Studies of captive rhesus macaques show that these monkeys won't react fearfully to a snake until they've first seen another monkey do so, after which they quickly acquire a genuine snake phobia. This phobia might seem to be a learned reaction, but other experiments in which one rhesus monkey is placed in a situation where another appears to be reacting fearfully to a flower will not create a flower phobia in the first monkey. Thus rhesus macaques appear to have an inherited, evolutionarily acquired potential for learning the snake phobia.[6]

But to call morality an evolutionary development doesn't really tell us much. We can't even be sure whether this third theory still limits morality to humans, and many evolutionary thinkers do believe that humans are decisively *The Moral Animal*, which is the title of science journalist Robert Wright's 1994 book on the subject.[7]

Morality evolved among humans only? The question is largely one of time. Those evolutionists who argue that morality is limited to humans are also limiting the period in which it appeared to some relatively recent moment.

Humans emerged as a distinctive species, *Homo sapiens*, only around two hundred thousand years ago, when, as the fossils show, they developed a generally modern anatomy and brain size. Is it the case that only *Homo sapiens* are capable of morality? Or perhaps morality somehow appeared somewhat earlier, with the evolutionary emergence of our ancestor *Homo erectus*, some 1.8 to 1.9 million years ago. Here was a clever creature who may well have tamed fire and begun cooking food.[8] Our ancestors were standing and walking upright even four to five million years ago, but *Homo erectus* is a species of savanna ape adapted to walk and run with a full modern gait and thus, apparently, a modern grace and fluidity. Indeed, this particular human ancestor shares with us enough physical resemblance that, magically transported to the streets of a large modern city, some individuals might hope to blend in—particularly if they're careful to wear a hat over that low and sloping forehead, which shelters a brain about two-thirds the size of the average contemporary person's.

So a humans-only theory of the evolution of morality might describe its beginnings among some rough versions of ourselves sitting around the fire and sharing a haunch of crudely cooked meat. Maybe it happened as long ago as 1.9 million years or as recently as two hundred thousand years ago, when our immediate ancestors—the start of the *Homo sapiens* line—developed the modern-size human brain.[9]

The idea that morality has evolved among humans only typically presumes that the critical element is neocortical brainpower, at least enough of it to ensure an ability to think rationally or to analyze. But if such high-level abilities are essential ingredients for morality, then it must surely have evolved even closer to the present time, since full human rationality and analysis seem to require complex symbolic language. Sign-language experiments with apes show that all our closest modern relatives—chimpanzees, bonobos, gorillas, and orangutans—can be taught to communicate relatively simple ideas symbolically.[10] Many other species will communicate specific if limited information through vocal signals, and many animals can learn to identify sounds with objects and concepts. Primatologists Dorothy Cheney and Robert Seyfarth tell the story of a Namibian pet baboon, Elvis, who assisted his human owner in car repairs. While the owner lay beneath his broken-down Land Rover, the baboon would assiduously respond to spoken requests for various tools and auto parts. Elvis sometimes had trouble distinguishing a number 12 spanner from a number 10, but he understood very well the instruction "No, the other spanner."[11] Rico, a border collie living in Germany, showed his comprehension of some two hundred different spoken words.[12] Alex, an African gray parrot, demonstrated through his own speech that he understood the words and concepts related to objects, colors, shapes, and numbers.[13] Well, the stories of smart, language-understanding animals abound, and wild animal communications can be impressively varied and complex.[14] Sign language of the sort the great apes can learn is certainly a form of language—but human language is still much more complex than any nonhuman communication, and the evolutionary emergence of human speech requires both the gross anatomy for speech production and the neuroanatomy sufficient to organize a limited series of physical acts in the mouth producing a limited set of sounds into the words and sentences that, structured grammatically, become the unlimited miracle of spoken language.

According to linguistic anthropologist Philip Lieberman, an anatomically modern vocal tract shaped to produce the entire range of sounds in modern speech is not evident in the fossil remains of early humans until around fifty thousand years ago,[15] which gives us one approximate date for the emergence of full language of the spoken variety.

Another way of thinking about the language date is to consider the history of the human FOXP2 gene, which is associated with, among other things, the production and comprehension of spoken language. Scientists first identified FOXP2 after they discovered a human family with several members who exhibited a complex disorder that included trouble following grammatical rules and comprehending complex sentences, and a lack of the motor control necessary to produce intelligible speech.[16] Members of the family with this disorder were also found to lack a normally functioning FOXP2 gene. A good homologue of the human FOXP2 can be found in other mammals—mice, for example. But for mice this gene is associated with some different events and abilities. Recent studies have shown that the disruption of the gene in mice causes "severe motor impairment, premature death, and an absence of ultrasonic vocalizations that are elicited when pups are removed from their mothers."[17] Mice and humans are separated by seventy-five million years of divergent evolution, and the mouse version of the gene is distinguished from the human version by three mutations, while the human and chimpanzee versions of the gene are distinguishable by two mutations. According to one team of geneticists, the distinctive mutations of this gene in humans producing our own modern variant of FOXP2, associated with the presence of some of the intellectual and a good deal of the motor ability required to speak and understand language, appeared at some point during the last ten thousand and one hundred thousand years.[18] That gives us a second reasonable time frame to think about the evolution of spoken language with the full modern range of expression.

Undoubtedly, language is necessary for moral debate. And at least one evolutionary psychologist, Marc Hauser, argues (in *Moral Minds*, 2006) that human morality emerged in a form strongly analogous to that of human language. Human morality, Hauser believes, amounts to a universal instinct, a complex system for recognizing and responding to *right* and *wrong* that has been evolutionarily embedded in the human brain. It thus powerfully resembles the universal instinct that linguist Noam Chomsky has long insisted characterizes human language. Certain aspects of the moral instinct may, Hauser argues, have appeared before the start of human evolution, but since it only flowered fully during our independent evolution, morality is essentially limited to humans, while animals may have something more like "pre-morality" or "proto-morality."[19] Just as Chomsky's universal linguistic grammar is limited to the human species, so Hauser concludes that his own universal moral grammar is largely a single-species phenomenon.

My argument in this book is different from Hauser's. Instead of focusing on evolutionary developments that may have happened only a few tens of thousands of years ago, I would prefer to imagine events more generally situated in the range of sixty-five million years ago (a time roughly marking the demise of dinosaurs and the rise of mammals) and even much earlier. I believe that morality predictably appears as an evolutionary response to group living, and we can therefore expect to see clear signs of full moral behavior appearing much, much earlier than the recent start of human evolution.

Yet why should group living have any evolutionary effect at all? How important is living in groups? The answer to that question can be considered, first, through a little introspection. Ask yourself whether, in your lifetime, you have ever experienced complete isolation for forty-eight hours. By *complete isolation* I mean not seeing or speaking to someone else for the entire time. How about twenty-four hours? Sixteen? We all have clear limits to our psychological tolerance for social isolation, and even moderate isolation, say traveling in a foreign country where we have no friends or acquaintances, creates after a few hours a low-level anxiety that, within a day or two or three, can turn into something more like desperation: a real sense of social hunger. We possess powerful emotions—some of them associated with the core feeling of loneliness—that emphatically prod us to seek the society of others.

Such an intense emotional need for it tells us that social interaction must have been an extremely powerful force shaping behavior during the evolutionary past—say, the last sixty-five million years. The need is not merely a case of looking for the joy and amusement that fellowship provides. We group-living mammals absolutely need others for our very survival, and thus others might be thought of as an environmental force as fundamental as gravity. Gravity is silent and invisible. We seldom think about it. Yet we can see that gravity has shaped us mammals profoundly. For the primates—prosimians, monkeys, apes, humans—gravity has had an additional evolutionary effect, since the primates are characterized by some extended history of living in the trees, where gravity works like a sharp knife. So we find some of the defining primate features—powerfully grasping hands and good binocular vision—serving as, among other things, important evolved responses to the problem of gravity in high places.

We fundamentally need others, and others can be imagined as an unforgiving force like gravity, a precarious environment like the trees. Depending upon how we behave around that force or environment, others will reward us powerfully or punish us profoundly. So why wouldn't we expect others to have deep and broad effects on how we behave and have evolved to behave around them? Others define the quality and nature of our lives, the comforts we can expect, the food we'll eat, the support and protection we'll get over a lifetime. Indeed, others often determine the timing and nature of our deaths. And since others include our mates and potential mates, others directly determine the fate of our own genes and account for 50 percent of our children's genes. No wonder our parents told us we needed to behave ourselves! What they really meant was that we needed to behave ourselves winningly around others.

Evolutionary theory provides a structural way to think about morality. That is to say, the evolutionary development of morality would have its own structure, based on how and why, where and when. That particular structure would, in turn, give form to the structures embedded through evolution into microanatomy—neurochemical systems associated with moral assessments and emotions—and from there to the structures of human or animal psychology. Each structure is shaped in some ways by its precursor. It's as if one forms a gentle, revealing layer over the other. Although the ultimate structures—evolutionary, neurochemical, psychological—are largely invisible, they are expressed more visibly by the proximate, overlaying structures of behavior and (for humans) language.

That idea is reminiscent of the wonderful image evoked by Captain Ahab in *Moby-Dick*, as he philosophizes in one of his Shakespearean rants about the meaning of life and the deep, natural malice embodied by a certain white whale. "All visible objects," Ahab declares, ". . . are but as pasteboard masks." Such visible masks are merely the surface reality of things, molded by the face of some greater truth lurking behind or beneath. To get at the invisible but truer "reasoning thing" behind the mask, Ahab would reach through that visible mask to probe and understand the invisible face behind it. Or rather, because he's an angry, violent man who believes he's on a metaphysical mission, he would "strike, strike through the mask!"

Let me do something of the same now, by summarizing in a few broad strokes what I believe are the invisible structures of moral evolution that produced the visible structures of morality as expressed in behavior and, for humans, language.

The Invisible Structures. As noted briefly at the end of chapter 1, I conceive of morality as consisting of two complexly intertwined large systems, what I identify as the systems of *rules* and *attachments*. Attachments morality can be imagined loosely as relying upon "Cooperation" and "Kindness," which are the titles and subjects of chapters 10 and 11.

Evolutionary biologists have been thinking seriously about cooperation and kindness for the last few decades, although they have often preferred to discuss such things under the name *altruism* and some other specialized terms. Altruism is the big problem because a first understanding of evolution does not seem to predict it. We can think of evolution as a process whereby diverse individuals (with, say, modest differences in beak size, tooth structure, brain size, or toe length) become sorted out in a fashion whereby those individuals with the most effective beaks or teeth or brains or toes are the winners: the ones who reproduce more successfully and thus more fully define the next generation. If we think that way, we are also seeing evolution as nothing but one big competitive struggle. Competition and cooperation may appear to be opposites, so how can a competitive evolution also produce cooperation?

Many people have tried to answer that question by creating a conceptual deus ex machina. Sometimes it's literally a god, one who fortuitously inserts cooperation into an otherwise viciously competitive world. Other times it's the idea of a cultural evolution that swims upstream against the powerful competitive currents of biological evolution to produce an opposite force among humans. But neither supernatural nor cultural interference can readily explain all the cooperation that we commonly see among animals of all sorts, and so we are left seeking the answer according to the principles of Darwinian evolution.

I see attachments morality as a completely explainable product of natural selection. We who live in groups do so in part because natural selection has brought us the means to cooperate and sometimes, particularly among mammals, the inclination to kindness. Natural selection can explain the appearance of altruism, which might be thought of as "the origin of goodness."[20] Yet the elegance of recent theoretical explanations for altruism and other aspects of attachments morality can easily entice us into concluding prematurely that morality is nothing more than attachments: a psychological system promoting, for example, cooperation and kindness, or altruism and empathy, or, in a political system, egalitarianism.[21] These are unsatisfactory conflations. If morality is essentially cooperation, do we then say that more cooperative people are more moral than less cooperative people? If altruism by itself explains morality, do we then say that people in the helping professions—doctors and teachers—are more moral than businessmen or car mechanics? This simplistic equation—morality equals attachments—is what I call the *niceness fallacy*: the false idea that nice emotions and nice behaviors explain morality.

The equation is not merely false, however. It is also biased. It expresses a common if unconscious bias in favor of the liberal Western perspective. After all, surely a Western liberal country such as Sweden is markedly more egalitarian than, say, Saudi Arabia or some other hierarchically organized nation. Any realistic look at human morality must consider the phenomenon in all its contemporary expressions, liberal to conservative, egalitarian to hierarchical, and everything in between. We should, as well, recognize that the highly authoritarian, rules-oriented morality expressed in the biblical Old Testament has as much a place in people's understanding of morality as does the vision of an empathy-based, egalitarian utopia promoted in the biblical New Testament.

Rules morality complexly intertwines with attachments morality. The attachment virtue of cooperation, for example, works best when people follow the rules. We have moral rules against deceit and theft, which are extremely important in sustaining cooperation. Try dealing with a dishonest plumber, as I have recently done, to see what I mean. But rules morality may in general have evolved not so much through natural selection as through social selection. *Natural selection* refers to situations in nature selecting out, generation after generation, the evolutionary winners. *Social selection* also operates as a Darwinian evolutionary process, and we might imagine it as one variety of natural selection, except that social situations are what methodically sort out the evolutionary winners from the losers over extended time.[22]

Social selection works not through some imaginary group consensus about how the ideal society ought to be constructed. Instead, it develops through the predictable interactions of individuals who may ultimately have their own self-interests at heart. Social selection occurs in a group through the negotiations between individuals with different levels of social power, individuals who can roughly be characterized as insiders and outsiders. In any group, there are going to be some socially powerful individuals; we might identify them as the *insiders*. The socially less powerful individuals might be called the *outsiders*.

Think of high school. You enter high school as a trembling, uncertain outsider, a mere freshman surrounded by socially more powerful individuals who indeed appear bigger, stronger, more self-confident, sexier, wiser. Those are the insiders, a few mature members of your own freshman class and many members of the higher grades. You the outsider feel a strong need to fit in because your emotional nature tells you that fitting in is important; and, of course, you can also see for yourself that fitting in is probably far more fun than not. Moreover, you recognize that fitting in with this group will be important for your future well-being, and you will soon discover plenty of psychological torments awaiting those who fail to fit in well. Moderate to severe ostracism. Teasing. Having no close friends to take your side in an argument or a fight. And so on. If we think of high school as a rough example of what social life for humans might have been like in the Pleistocene—life within a relatively small band or tribe of people intensely dependent on each other's tolerance and goodwill—we can see how the social dynamics of a group can powerfully affect how you behave. Fitting in with the group, responding to the approved styles and behaviors (as defined by the most socially powerful individuals in that group), becomes important. To be sure, in the Pleistocene the stakes were much higher, since instead of being left out of the prom you might have been left out of the hunt and the feast that followed. Ostracism in high school can be intensely painful, even psychologically damaging. Ostracism in a Pleistocene band could mean death by starvation or worse.

So far I have been describing a strictly social situation, one to which we respond through social learning. That process we sometimes call *socialization*. It should be clear that this kind of learning—the learning of manners and morals—is absolutely part of our social experience today, as it has always been as long as we and our ancestors have been living in groups. Insiders define the rules. Outsiders learn the rules and try to follow them, with the expectation of becoming, one day, insiders themselves. But such a social pressure, something we see commonly, can simultaneously work as an evolutionary pressure creating a type of evolutionary process—evolution through social selection—if it meets two criteria.

First, the pressure for someone to acquire the desired social behaviors should be intense or focused enough to affect reproductive success. Maybe that's not so likely in high school, since one of the great wonders of high school is that you can graduate, move on with your life, leave behind those painful years of datelessness and maladjusted malaise. You'll find a good mate and a good life after you graduate. In a Pleistocene band, though, you'd better fit in with the band you have, at least well enough that you're not shortchanged on food or left with no mating opportunities because no one trusts or likes you, or, one day, simply executed by a plotting group of your disapproving peers. Pressure of that kind can work in evolutionary terms, since it can affect your survival and reproductive success.

The second criterion for social selection to occur is consistency over extended time. The desired social behaviors should be much the same with each new generation. You might imagine that such long-term consistency is simply impossible, because you might have the false impression that desired social behaviors are all random inventions, quirky little behaviors spontaneously created by the socially powerful to promote themselves and hold back the rest. That may be true for some of the strange fads in dress and speech that sweep through American high school culture regularly. But in major realms of social contact the expectations of the socially powerful are going to be predictable or consistent over time.

If those two criteria—pressure affecting reproductive success, and great consistency over extended time—are met, then a social situation will produce not only social learning and adaptation but also a deeper kind of adaptation, an evolutionary kind where, over fifty or a hundred or a thousand generations, certain ways of behaving can become embedded in the neurochemistry of the brain, and thus the psychology, of individuals. How long does it take before social selection embeds itself in the psychology of individuals? It would depend upon the intensity and focus of the selection. It is worth noting, though, that a recent experiment in domesticating wild foxes took fewer than twenty generations to produce foxes who were not only physically very different from their wild counterparts but psychologically and cognitively different as well. The domesticated foxes would come when called by name, and they were able to understand the meaning of a person pointing at or gazing in the direction of a remote object. Wild foxes don't and apparently can't do that.[23]

Anthropologist Christopher Boehm speculates about a social selection that may have occurred among bands of early humans during the Pleistocene, perhaps as long ago as forty-five thousand years, and he does it through an analysis of social data on ten present-day groups of hunters and gatherers, people living in what are, by his assessment, "Pleistocene-appropriate" societies.[24] Boehm found the social norms in all ten societies include a preference for individuals who were cooperative, generous, and inclined to share. All ten societies express a normative disapproval for murder, theft, adultery, sorcery, and disregard of taboos, while most of them disapprove of bullying, violence, incest, cheating, lying, or dishonoring a female.[25] He also found that the ordinary response to violations of such norms runs from aloofness to ridicule, ostracism, physical punishment, and execution. Thus, we can imagine this system as consisting of a set of social rules with a series of possible sanctions to enforce the rules. It looks like a rather simple system for social learning, but Boehm's argument—and mine—is that such rules and enforcements can have an evolutionary effect over time, since consistently reducing the success of those individuals who are temperamentally more antisocial can amount to an evolutionary selection.

It is easy to imagine, as Boehm does, that such a social selection would be limited to humans, who can use their intelligence to think about such things and their language to gossip and plot. Yet recognizing a bully or a cheater or a murderer or an unreliable individual is not an intellectual process but an emotional one. You don't need to think or analyze or calculate. You just need to feel. A bully makes you angry. A cheater leaves you depressed. A murderer creates fear. An unreliable individual is one you feel like avoiding. Animals can experience these feelings, too—and at least sometimes, it would seem, surprisingly more complex ones. Brown capuchin monkeys from South America, so one recent laboratory experiment has shown, become upset in situations of unfairness such as when food treats are unequally distributed.[26]

Language creates the ability to gossip and plot, and thus it enables a significantly collectivized response to antisocial individuals. Execution is the ultimate way of dealing with political bullies, and according to Boehm's data, executions have still been reported "on a number of continents" among modern hunters and gatherers.[27] Yet just as high intelligence is not needed to recognize unpleasant or antisocial individuals, so language is not necessary for a collective response. We know this from the example of bonobos, an ape species where females regularly gang up to put out-of-control males in their place. The evolutionary effect of female group action among bonobos may, indeed, account for the fact that bonobo males are significantly less aggressive and less physically powerful than their chimpanzee cousins.

Boehm's theory can explain the social and evolutionary development of egalitarianism among early humans, just as it could explain the origins of egalitarian tendencies among some nonhuman species. But we should also recognize that egalitarianism is not the only way to handle destructive bullies and other antisocial individuals. A second approach would be to incorporate them into a dominance hierarchy, rather in the style of chimps or modern humans living in hierarchical societies. In a simple dominance hierarchy, a single individual at the top becomes the accepted leader: the alpha male or the king or the power-wielding high priest. A king and kingdom may not seem like the best solution to the problem of political bullies, but of course it depends on the king. A good king maintains order, punishes unfair acts among subordinates, promotes a general code of reasonable behavior, and thus creates a stable social order where everyone receives some level of protection from the random antisocial acts of others. A similar effect can be found in some animal societies that are hierarchical. Take pigs, for example. Placing a dominant boar in the pigpen reduces fighting—as Temple Grandin found experimentally. "When the younger pigs saw him coming," she writes, "they stopped fighting. It was exactly like a bunch of young hoodlums who see the police and instantly stop what they're doing." Too anthropomorphic for your taste? In fact, Grandin declares, "The younger pigs acted so much like young human males they would even look around to see where the boar policeman was *before* starting a fight."[28] Another researcher found that even spraying the odor of an adult boar into the pen reduces fighting.[29] Hierarchies can control bullies and reduce random bullying.

The Visible Structures. Morality, invisible at the level of evolutionary theory, becomes visible in behavior. We can daily witness moral behaviors expressing an underlying structure. We can see attachment behaviors—spontaneous acts of kindness based on a feeling of empathy, for example—that appear to display something we recognize as morality. We can also routinely see rules morality acted out—though sometimes most emphatically through the violation of rules. (The thief is more likely to remind us of a rule against theft than the honest person, perhaps partly because honesty is common enough in many places that we often take it for granted.)

Once we begin looking at morality in behavior, we discover universality. Beneath a relatively minor patina of cultural variation, we see that the morality of humans is fundamentally the same in every culture. That's one reason why you can travel to any place on the globe and find yourself treated with a surprising measure of decency and fairness, even when you don't understand the language and no one understands yours. You can still communicate crudely with gestures and pantomime, and that should be enough. If you can somehow express your needs, you can expect that most people will treat you there, in that exotic culture, in much the same way you would expect to be treated back home. True, you might accidentally stroll into the wrong part of town and fall victim to a person harboring malicious intent; but the same could happen back home. True, you might find a place where people are more than usually suspicious of outsiders; but you already know what suspicion of outsiders feels like, and you also know from experience that it can usually be overcome with time and patience. Yes, you might even have the bad luck to arrive in a war zone and be regarded as an enemy by the people around you. But you already know what wars are, and you already understand what the moral rules look like once people are divided into allies and enemies. Murder is murder everywhere. Theft is theft everywhere. Treason is treason everywhere. Just as, on the positive side of things, generosity, sympathy, and charity look the same no matter where you find them or what language you speak. These are all aspects of what appears to be a universal human morality, and they are marked by a universal series of moral rules expressed in a similar fashion around the world.

As we shift from looking at moral behavior to looking at the expressions of morality in language—written moral codes—we find once again evidence of a provocative universality. The Judeo-Christian Ten Commandments may be the written set of moral rules most Americans and Europeans are familiar with, but go to the Middle East or Asia or Africa and you'll find an impressively similar set of rules listed under some different title, written in a different language, with a different social and cultural history behind it. In Islam, for example, you can discover the Ten Commandments clearly echoed in the Koran, particularly among the precepts of wisdom found in "Al-Isra" ("The Night Journey").

If universality at the species level were the sole shape of morality, life would be a lot simpler and this chapter a few paragraphs shorter. Within that provocative universality, however, is a peculiar singularity: one of reference or focus. The human moral rules may look the same around the world, whether we see them acted out in the drama of behavior or whether we examine them as they're written down in the documents offered by various cultures—except that the rules ordinarily refer to particular social groups.

We see this referential quality in the Ten Commandments. As the biblical text explains in the book of Exodus (and, in slightly different form, in Deuteronomy), the Ten Commandments were handed down, along with a much larger set of instructions, as part of Yahweh's intended singular relationship with the people of Israel under the leadership of Moses. Altogether, traditional Judaism holds that there are 613 mitzvoth, or divinely given rules and instructions, serving as the canon of Jewish law, and the Ten Commandments can be regarded as a central part of that divine gift.

The cultural reference of the Ten Commandments has historically expanded, so that today many people regard them as not merely part of an ancient Judaic document but also of a Judeo-Christian one, something observed and revered by millions of people across the world. Nevertheless, even with that extraordinary expansion of reference, the Ten Commandments still refer to a particular social group—that of observant Jews and Christians.

That a moral code can be both universal in its general structure and yet, in its focus on a single social group, particularized may seem obvious or irrelevant. Yet it could account for the disturbing observation that suicide bombers—young men (and occasionally women) so driven by their internalized distribution of rules and attachments that they will kill themselves as well as countless others—are morally normal, or within the normal range, rather than randomly violent psychopaths. They are unfortunate young people who find that their own internal voices coincide with a few voices around them, and based upon an intense loyalty to one particular social or cultural group, they become moral warriors for the glory of God and the promise of heaven. Actually, we can find a similar pattern spread broadly outside the human story, and at least among a few species, it can emerge with an intense ferocity. Lions, hyenas, or chimpanzees, for example. This subject I'll develop more fully in chapter 6.

But language enables humans to formalize or concretize moral rules—to write them down and refer to them as if they're cut in stone—and that brings the possibility of reinforcing loyalties through formal systems and thereby expanding social group size. Indeed, a historical reading of the book of Exodus might well regard it as an account of the arrival of written Hebraic law necessary for a radical expansion of group identity.

We have here the story of twelve tribes coming together in a single camp at the base of Mt. Sinai, with Moses nominally their leader. Among the first problems Moses encounters is that of adjudication: resolving disputes. Moses tries to serve as a judge but is quickly overwhelmed with the work. His father-in-law, Jethro, chastises him: "You are not able to perform it alone." Jethro then advises his son-in-law to select honest men as judges, men "who hate a bribe," and to inform them what the laws are: "And you shall teach them the statutes and decisions, and make them know the way in which they must walk and what they must do" (Exodus 18:17–21). In the context of this historical need for a formalized set of laws, we can consider the arrival of the Ten Commandments and the rest of the mitzvoth.

Moses goes up to the mountain to commune with God, or Yahweh, and receive the laws. Trouble begins, however, once Moses comes down from the mountain with the Ten Commandments chiseled into stone and discovers the people of Israel already embroiled in disobedience, conflict, and rebellion. During his absence, and with his brother Aaron's acquiescence, many of the Israelites have fashioned a golden calf to worship, and now Moses, outraged at the sight of this new god, breaks the stone tablets. He seizes the golden calf, burns and pulverizes it, mixes the ashes into water, and forces the guilty Israelites to drink the full concoction. Moses next assembles all the sons of Levi—men of his own tribe from Israel's twelve tribes—and instructs them to move through the camp and, with their swords, punish the people for their waywardness. The Levites kill around three thousand that day, we can read, whereupon Moses tells them that they are now "ordained" for service to God (Exodus 32:29). Having broken the first set of stone tablets in anger and organized the killing of three thousand golden-calf worshippers, Moses finally returns to the mountaintop, where God a second time writes down the Ten Commandments on two stone tablets.

This story—the rebellious people, the punishments and killings—leads us to imagine an unstable social group in danger of breaking apart. We can also speculate about a possible power struggle between the two brothers, Moses and Aaron, both competing for the leadership of this new and larger nation by invoking the power of competing gods. Using the might of the sword and the chisel, Moses wins. Moses is able to get his laws written down and even have the essential core of them chiseled into stone as a lasting cultural document, which gives him the legitimacy he needs. A political reading of Exodus, then, would place Moses at the center of the action, and it suggests that Moses himself may have written the law and Commandments.

In any case, we can see that the Ten Commandments are part of some basic written rules important for people suddenly faced with the challenge of living in a newly enlarged social order. The Israelites might have had no need for written rules at all when they were living in smaller tribal societies that were, say, not radically larger than the size of a chimpanzee community. Such tribal groups would have been small enough that everyone recognized almost everyone else. There would have been little confusion about what belonged where and who could do what to whom. Anonymity and diversity would not have placed their complicating strains on the social order. Nor, when the Israelites were enslaved in Egypt, unhappy subjects to the greater rule of the pharoah and his Egyptian system, would they have needed written rules. But now that they are no longer subjects of the pharoah and Egyptian law, the people of Israel have become subjects of themselves; and along with that new independence comes a new group size. Instead of a few hundred members per tribe, there are now thousands, all living in a single camp and needing to function as a single nation. This new situation requires someone to fortify the moral rules essential to living in a group: to write them down in order to enforce them through a formalized system of priests, judges, courts, and penalties. That's how cultural developments can take over and modify evolutionary ones, but at the base of it is an evolutionary structure that includes a singularity in focus and reference on the particular social group.

Language has given our human moral nature a visible shape in the form of spoken and written moral rules. The visibility that language brought also produced the potential for more complex levels of social organization, so that humans, once banded like most other mammals into small social groups of a few to several dozen individuals, could eventually assemble into coherent nation-states of a billion or more. Yes, language and all its subsequent gifts (invention, teaching, culture, symbols, history, mythology, laws and contracts and constitutions) have enabled us to expand our repertoire of group size tremendously. But group size is not a measure of morality; and language has had relatively little effect on the essential appreciation of right and wrong among humans. The human moral structure is similar no matter how humans are grouped, whether they live as loyal members of a North American nation of three hundred million people or as members of a small, isolated, and politically independent village in the Amazonian rain forest of South America.

It's fair to ask whether rules can exist without language. They can. We pay attention to unwritten and unspoken rules all the time.

Rules are psychosocial systems for influencing behavior, and they have two essential parts: a reference to the behavior to be influenced and a reference to the likely enforcement. Neither part has to be written or spoken for a rule to work, and often neither is. Or maybe only the first part will be written, as in No Smoking, which includes the unwritten but understood likely enforcement clause *Or you could be fined or chastised.* Yet probably the great majority of rules remain invisible in both parts. They don't need to be written because they have become an aspect of our understanding of cause and effect in the social world, and over time we have simply absorbed them and thus grown up and learned to be graceful and obliging social beings. We may have learned these rules early in life—rules such as *Don't poke someone in the eye, or you could be poked back or scolded or otherwise embarrassed.*

We often imagine, in error, that rules and their enforcements are entirely social and cultural systems. Indeed, punishing a thief is a social act and ownership a social situation with cultural support. But the emotions that drive us to value ownership and to respond in anger or with outrage to theft are evolutionary products embedded in our psychology.

We often believe, wrongly, that rules enforcement must inevitably be organized and collective. Rules enforcement in many human societies has indeed become highly organized and methodical. But before written rules and laws, and before the development of official systems of enforcement, human moral rules—for example, the rules in favor of ownership and against theft—would have been enforced less formally. Whether enforcement is informal or formal, the potential thief is always restrained by the possibility of an enforcement against theft, which could appear in the form of a beating or shunning or some other punishment. So enforcement need not be organized, just as it doesn't have to be collective. It could be individual.

The psychosocial moral rules arise in the presence of serious social conflict. They serve to reduce such conflict, and so moral rules—written or unwritten—can be predicted because we can recognize where social conflict is likely to be most intense. I am imagining five areas, or *realms*, of social contact where the potential conflict is sufficient that the emergence of moral rules is predictable with or without the presence of language. These are the realms of authority, violence, sex, possession, and communication, each of which will become the subject of a chapter in Part II of this book.

In his comic film *History of the World, Part I*, Mel Brooks has Moses coming down from Mt. Sinai with Fifteen Commandments written on three tablets. Alas, the clumsy leader of the Israelites accidentally drops and breaks one of the tablets, leaving only two tablets with a mere Ten Commandments. It's an amusing if irreverent joke, and it reminds us that the fourteen to fifteen imperative statements actually appearing in Exodus 20 have to be squeezed, somewhat arbitrarily it seems, into ten. Moreover, if we examine the full text in its most general aspect, we can discover, following the thinking of St. Augustine, three essential themes. The early commandments define the proper relationship between God and humankind; the middle commandments identify proper relationships among people; and the final one clarifies the appropriate private thoughts for any individual person.[30] I divide the Ten Commandments into five essential rules, ones that I believe are valid for social animals generally because they arise in response to conflict within the five realms of social contact.

One of the glories of being human is our unrivaled ability to use complex symbolic language, which enables us to communicate quickly, glibly, and sometimes fluently about moral issues. We engage in earnest discussions among family members and friends, and with strangers over the radio and television and across the Internet. Often, these are moral arguments, ones in which the participants toss ideas back and forth with the hope of stimulating in others a moment of real emotional engagement concerning some moral issue. I try to persuade you that the relevant emotion for this situation is outrage at the injustice that has been perpetrated. You try to persuade me that the most important emotion is alarm at the violation of some particular rule. Now, however, a third person enters the conversation and attempts to provoke our intuitive sense of attachment, our empathy for unfortunate victims, while a fourth person warns us all to be clearheaded enough to consider ramifications that could ultimately lead to a collective failure in the authority of our leaders and institutions. We should feel anxious, he says. We should feel angry, you say. We should demand justice, I say. We should wish for fairness. We should experience empathy. We should feel guilty. We should follow our urge to punish others. . . . And so it goes. So the verbal discussion, our magnificent keyboard of words and sentences and language, reaches down into the mammalian parts of our brains and enables us to play an elaborate chorus of moral rules and moral attachments, and the various emotional voices and tones and chords they're associated with.

Our glorious ability to use words allows us to discuss morality and to debate, endlessly, this or that obscure issue about it, to write down rules and laws, to comment on and amend them, to think of new areas where old truths might be relevant. But our general system of morality is not particularly a product of our high intelligence or refined rationality or even our ability to manipulate words. We don't behave in good and bad ways because we listen to the folly or wisdom of our outer voices, but rather because we listen to the folly or wisdom of our inner voices: those desires and inclinations embedded, through ancient evolutionary processes, in the emotional parts of our brain. Those parts evolved into much of their present form long before the separate evolution of our own species. They are parts, therefore, that we share in common with a large number of other animals.

Language can help reveal the invisible structures of human morality. We can read the rules, talk about the problems and choices. But we can also recognize that our spoken and written words express unspoken and unwritten universes of urge and inclination and inhibition—the complex, underlying, invisible mental structures that together make up human moral psychology. Invisible worlds of a parallel kind exist among a large number of animals, but since they lack language, their underlying structures will be made visible mainly through behavior.

PART II

What Is Morality? The Rules

CHAPTER 5
Authority

"Whales in the sea
God's voice obey."

—Herman Melville, *Moby-Dick*[1]

When your dog does something that seems right or appropriate, you let the canny canine know unambiguously what the meaning of that act is. "Good dog!" you say, with all the emotional emphasis you can muster, reinforcing the verbal stroke with a physical one. When your poor pooch does something you regard as wrong or inappropriate, you let him or her know with finger-pointing, scowls, and other nonverbal discouragements as well as the powerful comment "Bad dog!" That's human language, of course, and your dog could be seriously bewildered by such gibberish and the antics accompanying it, but you are still trying to teach him or her to follow a set of rules, and in distinguishing *good* from *bad* for the benefit of yourself and the dog, you are using the language of morality. It's odd, when you think about it, to use moral language when talking to a dog, and it makes me wonder what this tells us about dogs.

Sure, these are only dogs I'm talking about. Good dogs and bad dogs seldom seriously affect our own welfare or impinge upon our own human moral systems. If we want to look for the moral behavior of dogs, we should really look at how dogs behave with one another. But dogs, in their intimate relationships with humans, present an interesting and possibly unique situation. Although I would argue that looking for signs of dog morality ultimately requires examining dog-on-dog behavior, their special relationship with humans and our unusually intimate knowledge of them provides a special opportunity, at the least a useful analogy or model to begin thinking about animal morality.

You and I already quietly acknowledge the existence of that model in the way we use *good* and *bad*—the language of morality—to describe much of dog behavior. In training our dogs, we are also actually trying to teach them a strange if telling version of the Ten Commandments—radically modified, of course, to account for the animal's size, physique, and special interests. *Don't kill cats. Refrain from taking things that aren't yours, such as the sliced cheese and crackers on the coffee table. Don't try to deceive* me *about who left that stinky puddle in the kitchen. And don't even* think *about going after the neighbor's estrous bitch. Follow those rules, and you're a good dog. Violate them, and you're a bad dog. . . .*

That's the code, the set of rules we humans would like to see our dogs learn. And yet, as every dog owner knows, all aspects of the code are secondary to the opening and primary rule, which is *Obey your master.* This instruction is so essential that when we take our canine to a real professional for help with the training, we don't say we are taking the dog to Training in the Basic Rules and Tricks of Being a Good Dog School. We say we are taking the dog to Obedience School.

I know about Obedience School because I've had experience with a disobedient dog. I'm a pragmatic pet owner, which means I believe in obedience as a matter of protecting a dog's well-being and, of course, my own and that of anyone else around the dog. I am not interested in having my pooches sit on command because I take pleasure in seeing them sit when I say, "Sit." Rather, I want to be sure they'll stop moving when a truck they may not be aware of is careening down the road.

The disobedient dog is named Smoke, and as I write these words, she lies sprawled before me, sacked out on the living room rug, her enormous black nose and snout resting between two giant paws, looking every bit as angelic as her somewhat larger sister and littermate, Spike, who is also asleep on the living room floor.

Littermates are trouble. That's what the obedience trainer said on our first day at Obedience School. It felt like getting bad news from the doctor. Littermates are trouble because they get distracted by each other and are less attentive to you, their owner and aspiring master. Smoke and Spike were still just puppies then. I would let them out in our yard, then call and try to get them to come with positive and exaggerated gestures and lots of wild whistling. They'd look at each other, as if momentarily pausing to ask their favorite playmate, *Shall we pay attention to him and stop playing?* The answer was obvious: *Let's keep playing!*

The trainer had a sensible solution to this problem: "You could always give one away. At this age, a dog can adjust quickly to a new home." I already knew which one to give away: Smoke. The bad one. Thinking about giving away Smoke, however, made me feel sad and sorry, and I couldn't bear the thought of never seeing Smoke again.

Smoke and Spike. You can see evidence of their sisterhood in the color of their hair, or fur, which is orangey on their bodies, moving to white patches on their feet and tails, and turning to black around their muzzles. Their mother was a medium-size Saint Bernard, their father probably, so it was said, a mixture of husky and chow. On good days, when they're feeling well and are a little tired, they walk side by side on the leashes like a team of horses, turning their heads in unison to look or sniff at something interesting, and even, occasionally, wagging their tails in unison, which produces a dual-pendulum effect. Once, when I was walking them in tandem, a young child looked at us, gasped, then laughed and said to me, "I thought you had a two-headed dog!"

When people ask me who is Smoke and who Spike, I usually give them this simple mnemonic: Smoke is the one with the fluffy or smoky-seeming coat, whereas Spike is the short-haired one with the spiky coat.

A lot of people think Smoke is female and Spike male because, first of all, Spike is taller and about ten pounds heavier. She's solid and barrel-chested. If you didn't understand her inherent sweetness, Spike would look like good material for a guard dog. Spike seems serious and steady, whereas Smoke seems more a flighty dilettante. And Smoke is prettier than Spike. Smoke has a perky face, a sweet habit of cocking her head when listening to a new sound, a prancing walk, an enthusiastic bounce when running. Smoke is animated, always on the lookout for some new excitement. Her triangular ears are often half-raised, with the top ends folded down, flipping and shivering like ineffectual wings as she walks. Spike is longer-faced, slower, and more sedate. Her big ears just droop and flop like overcooked cabbage leaves.

Since Spike is bigger and generally more direct and imposing than Smoke, you might think that Spike would be the dominant one. In fact, I've never been quite sure who dominates whom. Smoke, for all her petite prancing enthusiasm and sensitivity, is also agile and athletic. Until they reached later middle age, Smoke could easily leap right over Spike. They would play, running in circles, each trying to grab the other's tail in her teeth and pull, and then, sometimes, they would run off again, turn around, and dash headlong at each other—with Smoke simply lifting her legs and sailing clean over Spike at the last minute.

Smoke and Spike graduated from roly-poly puppyhood and enrolled in active adolescence, and still following the teachings of the Obedience School trainer, I continued working with them, mostly on a leash. I taught them both to sit on command, to lie down, to heel. On leash, they were fine. Off leash was the problem. I remember one time letting them loose in the one-acre woods near my house. No other dogs. No cars. Just trees, bushes, and rocks. What could go wrong? After a few seconds of freedom, Smoke bounded off. I called and whistled, called and whistled. Spike soon arrived and sat down obediently before me, ready for the leash, but where was Smoke? Smoke had disappeared—shot like a four-legged projectile, as I eventually concluded, through the woods and out the other side, then down a street and across a dozen backyards. I found her a half hour later in another neighborhood, following some invisible trail and at least worn-out enough not to run away when I approached.

Then there was the park: a large piece of open land where, if you come before the town dogcatcher finishes his coffee at about 7:10, you can let your dog loose. As long as you clean up after the beast, no one seems to care. Half a dozen to a dozen dog walkers can also be found there at that promising time of the morning, standing in a circle and socializing while their dogs run in tangents and also socialize. It's a good place to bring your pet, and unlike the woods, the park is open enough that you can usually keep an eye on your dog or dogs.

I began taking Smoke and Spike there early in the morning. When I let them off the leash, they would chase each other around in large circles until they had released that first burst of enthusiasm and would then be relaxed enough to enjoy the random and sniffy exploration that comes with freedom. Freedom! So I would unleash them both upon arriving at the park, watch as they galloped wildly before moving off more sedately. Obviously, all this freedom was possible only if they were smart enough not to wander too far afield and obedient enough to remain ultimately under my control. Indeed, for about a year the park excursions worked . . . more or less.

Meanwhile, though, Smoke was slowly refining her vision of my zone of tolerance, and she had begun to observe that the zone was mainly geographical. She would wander closer and closer to the edges of the park, innocently examining the tufts of grass, and then—I can see it now!—she would raise her head, as if a squirrel or a squirrelly idea had suddenly seized her attention, and then she was off, bounding over someone's fence in an instant and scrambling away.

I tried everything. Long leash. New treats. Increasingly enthusiastic responses when she did come. And when she had taken off, left the park, and I managed to bound over the same fence and catch her, look out! She would get a scolding that must have left a memorable impression.

The last time I let Smoke off the leash in the park was not much different from several previous times. I had given both dogs their freedom, and they began with their usual excited running and mutual chasing. After that first burst of energy, they played some more, tentatively greeted some other dogs in the park, and so on. After about fifteen minutes, it was time to leash back up and walk home. I whistled, saw them both look. I whistled again. Spike began loping in my direction, while Smoke returned to her sniffy exploration. Spike came near. "Sit," I said. She sat. "Lie down," I said. She lay down. I snapped the leash on, and she was ready.

And Smoke? Smoke was now slowly headed in the direction of the street on the far side of the park. I called. I whistled once, twice, thrice. I began walking in her direction. I called again, "Smoke!" And then, "Sit!"

Was the dog hard of hearing? I whistled and called again, and at last she looked my way. Then she turned around and, in the proper style of her obedient avatar, began prancing in my direction. I could see she had something in her mouth, a prize of some sort, and she came closer and closer—close enough to look me in the eyes, and soon close enough for me to reach her collar. Almost. But as I leaned down to grab it, she bowed with her front legs, as if preparing to play, then bolted. But not far. She wanted to play. She wanted to tease. She began circling. She moved closer once more. Moving almost within reaching—or snatching—range. But the minute I made a slight movement in her direction, she bolted once again. Then circled. She still had that prize in her mouth, and she now began teasing Spike with it as well—approaching, retreating, circling, dancing playfully. This approaching and retreating may sound like ambivalence, but her manner was all about play: play of the teasing kind.

I've had dogs all my life. I've also raised two children. So let's just imagine that I have learned a few things about how to be an effective master for dogs and a good father for children. I believe in reason and consistency, and I know very well not to confuse anger with instruction. But the truth is, just then, at that moment, I was seething. Smoke was a *bad dog.*

Or was she?

What happened next was not something deliberate or calculated or sudden. It was more like a directional shift in my thinking. I began to imagine that Smoke wasn't a bad dog. I began to consider, instead, that maybe I was a bad master, that my dog was failing to learn not because of some fault of hers but rather through some fault of mine. So what was I doing wrong?

My Obedience School trainer had talked about the need for consistency—but I wasn't paying much attention because I already knew this one. You can't expect to train or teach a dog anything if you change your mind from minute to minute or day to day about what's important. That's Psych 101 in the theory of animal training. Be consistent. Be as consistent as a machine doling out food pellets to a pigeon. It's a good idea, but still, as I eventually figured out, a limited one. That particular training model fails to consider that even though *you* might be able to imitate a machine doling out treats, your animal is not a machine; and what I had failed to appreciate in this case were the unmachinelike differences in personality between Smoke and Spike. What worked just fine with Spike did not work so well with Smoke. I was being consistent, all right, but consistent to the point of rigidity. So the first and obvious lesson I learned here was this: Be consistent, but do so in a way that respects an individual animal's unique personality and individual needs, what we might call the animal's psychological integrity. Let me call this first principle *Show responsive consistency.*

My Obedience School trainer had stressed a second principle in dog training, which is the importance of showing dominance. The trainer repeated the theory that dogs are pack animals and therefore need a pack leader, an alpha dog. His idea was to remind Smoke in various ways that I was Alpha. For example, don't let her on the bed or on the furniture. Not an issue, because I already didn't. And, the trainer said, when teaching Smoke to walk on a leash, I should not move out of the way to avoid a collision with her. Rather, she should be given the opportunity to avoid colliding with me. Well, being Alpha Dog sounded good to me. Maybe I could put it on my résumé. The theory is repeated by a number of experts on dog training—including Oprah's favorite dog trainer, Cesar Millan, who declares in his bestselling book *Cesar's Way* (2006) that "the concept of the 'pack' is ingrained in your dog's DNA. In a pack there are only two roles: the role of leader and the role of follower. If you don't become your dog's pack leader, he will assume that role and try to dominate you."[2]

That sounds a little dire, doesn't it? Would my dogs let me go to the bathroom when I needed to? In any case, as animal scientist Temple Grandin points out in *Animals Make Us Human* (2009), the dogs-as-pack-animals theory is probably wrong. Grandin bases this assertion in part on the thirteen-year study of wolves done by L. David Mech on Ellesmere Island in Canada.[3] We know that dogs are descended from wolves. They're genetically wolves, more or less, and it stands to reason that the social structure of wolves should tell us something about the psychological needs of dogs. But wolves, according to Mech's study, are simply not pack animals—at least not in the sense of *pack* as a group of unrelated adults who fight it out among themselves to find out who's dominant over whom. Instead, wolves live in relatively simple family units—mother, father, and offspring—with the occasional adult stranger or adult aunt and uncle sometimes thrown into the mix. This doesn't mean that wolves live in egalitarian societies. It only means that no one has to debate dominance through threats and fights and politicking, or through demonstrating who can and who can't climb into the bed or sit on the furniture. In a family group, the structure of power is already established by the simple facts of family life: with the big adults automatically dominant over their small offspring, and the older siblings steadily dominant over younger ones. Mech believes the male wolves could be slightly dominant over females. He notes both mother and father of a wolf family hunt cooperatively, and they also eat the meat together, without any obvious complaints from the male. A mother wolf usually establishes a subordinate posture when meeting the male after a separation, although, as Grandin notes, this act gets the male to transfer to the female any food he may be carrying, either in his mouth or his gullet. That's food for the pups, and perhaps the mother's subordinate posture in this case is just normal begging for food.[4]

If I were to summarize my Obedience School trainer's second principle, I would say it means to assert dominance. But I might add, considering the new information from Mech and Grandin, that one should assert it not in the style of an alpha who has to prove himself but rather in the style of a dad who doesn't.

The third principle the trainer had talked about seems to fit in with the second, but it was also the idea that made the most sense to me. I understood it immediately because the trainer began by applying it on me. Out of the blue, he yelled at me, *"Get out of that chair and stand up straight right this very moment!"* I did not, partly because his yelling so distracted me that I never fully absorbed what he had said. He then repeated the same instruction in a normal, steady, confident voice. I understood the message the second time. This general idea is also espoused by Cesar Millan, who insists that the best emotional stance for a dog owner to assume is one showing "calm-assertive energy," to which the dog should naturally respond with "calm-submissive energy."[5]

Millan has been featured on television as "the dog whisperer," and I think his *calm energy* approach with dogs must be similar to Mae Noell's approach to apes. With her husband, Bob, Mae Noell for years ran an itinerant carnival show in the southern United States that featured boxing chimpanzees who would take on any brave human volunteers. The volunteers, tough guys eager to demonstrate their toughness, would pay a small fee, be fitted out with boxing shorts and padded gloves, then placed in the ring with an adult chimp also dressed up like a boxer and wearing padded gloves. Chimpanzees are quicker and far stronger than people, so it was a good show, although the chimp would end it suddenly by pulling down the volunteer's shorts. By the time I met Mae Noell, she was a widow running a roadside zoo in Florida called Noell's Ark Chimp Farm. The zoo featured the world's largest private collection of chimpanzees and gorillas, as well as a few other exotic animals. Her apes sometimes escaped, and I asked her how she got them back in the cage after that first glorious taste of freedom. The chimps might have weighed a hundred pounds or more, the gorillas up to three hundred or more, but any of them was strong enough to rip her arm off. How did she persuade them to get back into the cage? Oh, she told me, she would speak to them in baby talk. When you use baby talk, she explained, employing that singsongy, high-pitched voice, it's impossible to scare or threaten an animal, and so if they know you, they might listen. I believe that Mae Noell's baby-talk approach to apes is rather like Cesar Millan's calm-assertive energy concept with dogs. Maybe I should call this third principle *Don't confuse your dog by startling him or her, or by creating irrelevant emotions such as anxiety or fear.* Or, more simply, let me call it the *Don't get mad and yell at your dog* principle.

If this were a book on dog training, I might end this cautionary tale about my problems with Smoke by repeating the three principles I learned—responsive consistency, daddy-style dominance, and don't yell—and then telling you how well they worked. But I think I can summarize the problem and solution even more than that. The problem was that Smoke and I had entered an unhappy cycle where her disobedience made me angry, and my anger made her fearful and therefore inclined to avoid me in ways I considered disobedient. To get her to obey, I had to stop making her afraid. Spike needed guidance, rules, straightforward authority. Smoke probably needed those things, too, but she needed reassurance even more. I began paying closer and gentler attention to Smoke: a brand-new regime with lots of face-to-face talking and eye contact, a good deal of attentive stroking, and little to no scowly scolding.

And it worked! Or so I think. It's several years later now, and the dogs are discovering the benefits and limitations of maturity. Smoke no longer jumps over Spike, but she's also no longer a bad dog. She is still flighty enough not to be trusted in town off leash, but when I take them both out to a certain spot in the country, I know they can be released at one part of the dirt road, and then—after a quarter hour of rustling in the undergrowth, furious barking in one direction and then in another, more rustling in the undergrowth—we'll all rendezvous at a later place on the road. Same place every time. Smoke usually shows up first, and when I tell her to sit, she sits, waiting attentively for the leash to be put on. Spike now is the more recalcitrant of the two, but she will also finally lope out of the woods, sit down on the dirt road, and wait until her leash is snapped into place. Then both of them, with what looks like complete satisfaction—no regrets—continue walking, side by side now, along the country road. They wag their tails, and they seem to experience a peculiar bliss at being under my control. With their master at the other end of the line, these dogs look relaxed and secure.

Sure, dogs always enjoy the pleasures of freedom, which might be defined as the ability to go where your nose points. But dogs also enjoy, and need, authority. What dog is more lonely, distracted, and distressed than the ragged one over there without a master? So here, in this model situation with dogs, we find a dynamic relationship between the need for freedom and the need for authority. Not all masters are good at being an authority, of course. And an authority, to be most effective and appealing, should be fully appropriate. The authority should be what I call a *proper authority.*

The story of Smoke is a lasting reminder for me that an authority—a dog's master—can be imperfect or just plain bad. Being a good master or a proper authority, then, is not merely a matter of asserting power, but of asserting it in a way that respects the integrity of the individual. Dogs need masters. They need authority. They are psychologically built for it, just as we are. But they respond best to authority that acts or feels legitimate. They need a master who doesn't confuse them or make them anxious, a master who is able to promote a deep trust that over time translates into real loyalty.

The old cliché that Obedience School is not for the dogs but for their human owners is true. Dogs already know about obedience. They're born with a strong inclination to obey. It's the dog owner who needs obedience training, who must be schooled in evoking obedience. But if dogs are already inclined to obey, where did that innate inclination come from? Was it bred into the animal during dog domestication? Or was it already there in the original wolves who were transformed into dogs?

My answer would be no to the first, yes to the second. Wolves also already know about obedience. The inclination is already there, although it would never occur to them to obey a *person*. Like most wild animals, wolves are extremely shy. They fear people. They're obedient to each other.

Animal scientists like to speak about the authority relationships among animals in terms of *dominance* and *submission*, which recognizes an unequal distribution of social power. This unequal distribution of power manifests itself most obviously in the distribution of resources. Two monkeys walking in the same direction simultaneously discover a desirable piece of fruit at an equal distance between them. Which monkey will predictably get the fruit, and which will predictably wait for the leftovers? They won't usually fight over the fruit. That would be a waste of time, energy, and wellness, since one or both might be injured by fighting over a mere piece of food. Thus nature, by which I mean evolution, has given them the ability to respond intelligently to an already established dominance-submission system. Both monkeys already know who is dominant and who is subordinate or submissive. They already know who has a right to take that fruit—but just to clarify the situation, they may also exchange dominant and submissive signals to each other.

The signals of dominance and submission are generally universal among mammals, and that's one reason why you or I can often understand, just by looking at this behavior in an animal, what it means. Dominance and submission signals are another mammalian homologue, I believe, a set of behaviors similar because of a common origin, and they even work when an individual of one species (for example, a dog such as Spike) expresses submission to an individual of another species (for example, a person such as me).

Spike is emphatic about this sort of thing. When she wants some favor, such as to go outside, she'll approach me quietly, respectfully it could seem, and she sits down facing me. If I'm occupied with a book or poking at a keyboard on my lap, well, that's all right. She can wait. She waits with what seems to be endless patience, not moving, emitting neither squeak nor whimper, just gazing at me with what I can imagine to be a quietly supplicatory expression on her face. In such moments, Spike signals submission by both alignment and posture: an alignment clearly pointing to me, the master, and a lowered posture of patient submission. I think of this as Spike's prayer, a universal signal that I am the biggest authority around. In her circumscribed world, I am the Supreme Being.

As Temple Grandin reminds us, mammalian societies can express the distribution of social power in at least two broadly different ways: through a dominance hierarchy and through a family group. The dominance hierarchy may include several unrelated adults who have to figure out who is dominant over whom, and they may do that through displays, threats, fights, and politics (forming alliances). The family group would consist primarily of close relatives from one or more generations.

On first glance, it might seem that authority is only relevant in the society run as a dominance hierarchy, but that's not the case. As noted earlier, a family group also has a system of authority, with the adults dominant over the youngsters and, typically, older siblings over younger ones.[6] The major difference here is that the family group's authority structure is relatively stable. There are few fights over authority because the parents never lose dominance over the youngsters, even when the youngsters are grown-up. Grandin's argument about why dogs are naturally obedient to people and will (in ordinary circumstances) never try to dominate a human master has two parts. First, she notes that the dominance-submission relationship among dogs' closest relatives, wolves, who live in family groups, is stable. Grandin's second point recalls that scientists believe the domestication of wolves into dogs involved a genetic transition that turned off some important parts of the maturation process. In other words, dogs are forever immature wolves, and they show this paedomorphism in a number of ways, including floppy ears, curled tails, smaller brains, and a reduced fear of potential predators—including people.[7] Dogs are Peter Pan wolves. They are wolves who never grow up and therefore never acquire the normal fear-of-predators system that adult wolves acquire.

Scientists are reasonably sure about how dogs were domesticated in large part because they have a good example in the experimental domestication of Russian silver foxes, which happened at a large fur farm in Novosibirsk, Siberia. As you can imagine, foxes are ordinarily hard to handle because they're shy. They're innately frightened of people, who could after all be nefarious predators intending to kill them and strip off their skin. They'll panic at the sight of a person, perhaps try to escape through a far wall in their cage or enclosure or, if cornered, attack and bite.

In 1959, a geneticist named Dmitri Belyaev decided to try domesticating foxes on the Novosibirsk farm so that they would be easier to handle, and he did it in an extremely methodical fashion. He chose for breeding those foxes who showed less fear at the sight of a person—measuring fear at first solely by the distance people could approach before the animal tried to flee combined with the distance of the flight. Using that single criterion, a simple measurement for fear of people, and choosing the least fearful foxes for generation after generation, Belyaev created within about eighteen fox generations a domesticated species. They were descended from wild foxes but were quite tame. In fact, they were a lot like dogs. Their ears were floppy. Their tails curled. Their brains were smaller than the brains of their wild counterparts, and their faces were shorter and broader. Like dogs, their fur coloring developed into various piebald patterns, splashed with patches of white. These new foxes even barked like dogs. Even more remarkably, their personalities and cognitive abilities became rather doglike. As puppies they would cuddle up to people, solicit having their bellies stroked, and they would come when their names were called. Dogs are remarkable among animals because they will follow a person's gaze and pointing direction. Someone points at something meaningfully, and the dog understands. That's an unusual cognitive ability for an animal, yet these newly domesticated foxes had that same ability.[8]

Genetic mapping shows that Belyaev's domesticated foxes differ from the wild ones in forty different genes, so the experiment seems to indicate that selecting for a single trait—less fear—is likely to produce a suite of genetic changes.[9] Most of the newly acquired features are paedomorphic or immature ones, indicating that the domesticated foxes were, as dogs are, forever stuck in an immature state. Thus, it could be that nature's most direct route for reducing fear of predators is to turn down some complex genetic rheostat associated with the development of maturity. And because in his domestication experiment Belyaev so assiduously selected to reduce fear of humans, it reminds us once more that domestication is not a matter of increasing obedience so much as it is of decreasing fear.

It also reminds me of my problem with Smoke. I had trouble because, so I believe, my intensity made her fearful. But this principle—domestication through reducing fear—also suggests a way of predicting which wild animals people might be able to train most readily. Fear of people is almost part of the definition of *wild*, but some wild animals are innately more fearful than others, so if we are interested in capturing and training a wild animal—rather than working with a domesticated one—we might start with those animals who have no reason to fear most predators and would therefore not have evolved high levels of fearfulness. That might narrow it down to, say, lions, tigers, and elephants. Naturally, we would also want to work only with wild animals who are not likely to prey on us, which leaves elephants.

As wild animals who might be trained to do useful things, elephants are actually very promising. They're smart enough to learn quickly. They're big and powerful enough to perform acts people and other animals cannot. They don't eat meat, so they can be trusted not to eat people. And they have not evolved to be shy, in the manner of wolves or foxes, because they have no significant predators to fear (or at least didn't until recently, when powerful rifles and then automatic weapons were invented). Even lions get out of the way of elephants. Elephants seem so promising an animal to tame and exploit, you might wonder why no one has domesticated them, as people did by turning wolves into dogs and as Belyaev did with the foxes. I suspect that no one has produced a domesticated species of elephant because they're so huge and because their life span is about the same as a person's. These factors would make breeding for domestication logistically difficult. But perhaps no one has produced a domesticated species of elephant largely because there has never been a good reason to. Wild elephants will do just about as well. You just have to capture, tame, and train them.

People of the Indus Valley civilization on the Indian subcontinent figured out how to capture, tame, and train wild elephants four thousand years ago, and they used their trained wild elephants for farming, transportation, building, and warfare. Along with their trainers and masters (Indian mahouts), Asian elephants were first imported into the Mediterranean

Basin in Aristotle's time, after Aristotle's erstwhile student Alexander the Great acquired a large number of war elephants during his Asian campaign. Trained elephants thus entered the West as four-legged battle tanks, for which they proved supremely useful during the next few hundred years. They had one major weakness. If someone or something made them really afraid during battle, their obedience would collapse in a fearful panic, whereupon they would turn back and often cause enormous damage among their own troops. That's why, so some Roman historians claim, Hasdrubal of Carthage (Hannibal's brother) equipped his mahouts with a hammer and a long spike. If the elephant started to panic, the mahouts were supposed to drive that spike into the spine right behind the ears, thus killing the animal and avoiding the potential disaster a panicked and retreating elephant will cause.[10] Gunpowder and steel eventually made war elephants obsolete, but wild elephants are still captured and trained in the forested mountains of Myanmar (Burma) for the timber business.

A few years ago, I rode elephant-back for two days into the mountains of northwestern Myanmar, out to a riverside encampment of split-bamboo huts and a small shrine for U Dae, the *nat* (spirit) protecting elephants and their handlers. This was a temporary camp for elephants and their mahouts working at timber extraction in Plot 44 of Pantaung Township. I slept in a hut there and, during the day, witnessed firsthand why elephants are still used in logging. Elephants are more agile and versatile than most large machines. That elephants are smart is also an advantage because they can make quick, independent decisions: step out of the way in an instant, for example, in response to the peculiar sound of a giant log accelerating too rapidly on the downhill slide behind them.

The main reason I had gone there, however, was to learn about elephant training. I was curious: How does anyone persuade a wild elephant, who can weigh fifty to a hundred times more than a person, to obey and work for people? After leaving the extraction camp at Plot 44, I stayed for a time in the central elephant-maintenance village of Lay Te, where I met and chatted about training with the head mahout, U Tun Nyan, then had a long evening conversation with a former veterinarian, Dr. Soe Minn Aung. Upon my urging, Dr. Aung described elephant capture, taming, and training; and the following account is based on his comments supplemented with information from U Toke Gale's 1974 classic, *Burmese Timber Elephant*.

Wild elephants can be captured in a number of ways, but many people driving a wild herd into a large and solidly constructed corral is apparently still the most common. Once you have isolated a few younger wild elephants from this corralled herd, what then? First, the trainers use already trained elephants to push and pull and drag with ropes the struggling, screaming, wildly resistant young animal into a cradle, which amounts to a giant tripod made of thick poles and at least one standing tree. The cradle includes an overhead beam, and the elephant is immobilized with thick ropes tying his or her legs to the upright posts with additional ropes looped around the overhead beam and wrapped around the elephant's torso. This initial immobilization begins the process of demonstrating a radically new power relationship. The wild elephant is now powerless, unable to move, and people continue to break down his or her resistance by withholding food and water for two or three days.

Next, a pair of skilled trainers, working in shifts, begin bringing the animal food and water, but sparingly at first. Meanwhile, the two men chat gently and lovingly in a special language—the Khamti language, according to Aung—from five in the morning until eleven at night. As part of their strategy for keeping the elephant weak, the trainers make sure he or she doesn't sleep, and as much as possible, they continuously touch the animal all over, on the stomach, the legs, the chest, the rear, and even while hanging from a beam at the top of the cradle, standing briefly on the creature's back.

We could say the trainers are demonstrating their power to the elephant. They can do the essential things, such as bringing food and water to a hungry and thirsty animal. But the trainers also apply their power with some coherence and responsiveness to the individual elephant. They are consistent in bringing food, and they also use it as part of a larger system of reward and punishment. First, they test the foods—such as bamboo leaves, banana leaves, wild rice, tamarind-and-salt balls—to see what this particular elephant really likes, and then they feed him or her accordingly. Favorite foods reward good behavior. But if the animal continues to be rebellious, fighting, or angry, they cut the feeding altogether. Oh, yes, it must be harrowing for the elephant. U Toke Gale reports that in rare cases elephants have been known to commit suicide during this time, stepping onto the ends of their own trunks, cutting off their air and, hence, their lives. The elephant is "so bent on suicide that no amount of shouting, swearing, and spearing with sharpened sticks of bamboo could, in any way, scare him into removing his foot or relaxing its pressure on the trunk."[11] Ordinarily, though, these animals are quick enough to recognize that they have entered a new life and adaptable enough to accept their new tasks, new foods, and new leader. In fact, they have already been introduced to their new leader. One of the two trainers will become the elephant's mahout, a relationship that begins during training with the steady introduction of his distinctive voice, unique smell, and particular style of touching and will continue for the next twenty, thirty, or forty years.

It is certainly possible to overromanticize the relationship between a mahout and his elephant, but I did observe that the mahouts at Plot 44 extraction camp and Lay Te Camp would speak and sing to their charges, and they brought them down to the river for a good bath every day and even scrubbed behind their ears.

After ten days to three weeks of such treatment—controlled and limited feeding, induced sleeplessness, and the constant touching and talking—the young elephant is subdued and ready for more particularized obedience training. He or she is released from the cradle, hobbled at the forelegs with a length of cane, and over the next three months taught some of the standard commands and the responses required of a working elephant: *hmet* (lie down), *how* (be careful), *myauk* (lift), *pway* (carry), and *ya* (be quiet). So the new life of having a wooden bell strung around your neck, sometimes a chain wrapped around your foot, and of obeying the commands of your mahout in exchange for free food, a fair routine, new elephant companions, and a good daily bath and ear-wash, begins.

You might imagine that elephants can be captured and made to work for people because they're stupid or just inert. But one could just as easily conclude that elephants are smart enough to recognize quickly that they have no other choice, and so they intelligently adapt to the new life and new master. I think of the process as comparable to the Stockholm syndrome, which happens among people who are kidnapped and begin to identify with their kidnappers. And I believe that elephants learn obedience because they're psychologically already prepared to. Obedience to authority is part of the inherited psychological nature of an elephant.

This is true even though elephants live in both kinds of mammalian social systems I spoke of earlier: the family group and the dominance hierarchy. Actually, the females live in family groups, while the males leave their birth families around the start of adolescence to form bachelor groups composed of unrelated males, mostly, who sort out their relationships with one another by forming a dominance hierarchy. They develop this hierarchy through a good deal of testing and wrestling and even, occasionally, fights, but because it creates a system of authority within the bachelor world, it serves the socially positive function of reducing violence. Two elephant bulls aroused by the distant scents of an estrous female will, most of the time, already know who is dominant and who subordinate, so they don't have to fight over mating access. In the all-female family groups, meanwhile, authority is distributed more automatically, and typically the biggest and oldest adult female will be the group's highest authority: the matriarch. Any observer of elephants can sort out the distribution of social power by observing the signals of alignment and posture—alignment particularly in the all-female family group, where the matriarch leads and the others line up in single file to follow.

Authority begins with power, but mere power without a reciprocal respect for the integrity of the subject can inspire outward obedience unsustained by inward loyalty. It's the case with people, and I suspect it must be the case with dogs and elephants and any number of other animal types. Otherwise, all masters and mahouts would be equally good ones, and Hitler would have been indistinguishable from Churchill.

It is commonplace to believe that obedience is the very enemy of positive morality. After all, you and I can easily point to historical instances of people who have done unspeakably heinous acts and then proclaimed their innocence with the defense of obedience. They were, they plead from the prisoner's dock, just following orders. Excuses of obedience, the defense of following orders, recall for us the shameful cowardice of Nazi war criminals, and we are appalled at the idea of someone parading innocence on the back of obedience. The just-following-orders defense is seldom convincing because we know intuitively that obedience by itself is not enough. We understand that obedience as a moral theme always requires an assessment of the source or object of one's obedience: the nature of the authority one is being obedient to. So when we speak of *obedience* as a moral virtue, we actually mean *obedience to proper authority.*

Still, even obedience to proper authority may seem like a minor or secondary moral virtue. Not in itself so very significant. Hence it may be surprising when, upon returning to the biblical Ten Commandments as our working example of a moral code, we find that the Commandments actually open on the theme of obedience to proper authority. I quote from the stirring first lines of Exodus 20:

> And God spoke all these words, saying, "I am the Lord your God, who brought you out of the land of Egypt, out of the house of bondage. You shall have no other gods before me. You shall not make for yourself a graven image, or another likeness of anything that is in heaven above, or that is in the earth beneath, or that is in the water under the earth; you shall not bow down to them or serve them; for I the Lord your God am a jealous God, visiting the iniquity of the fathers upon the children of the third and the fourth generation of those who hate me, but showing steadfast love to thousands of those who love me and keep my commandments."

These initial verses are regarded, in most Protestant Christian traditions, as consisting of a preface ("I am the Lord your God") plus the first two Commandments, a pair of interrelated prohibitions against worshipping gods other than the true God and producing images to worship as if they were gods other than the true God. They powerfully eschew obedience to improper authority and, conversely, command obedience to the God who revealed himself to Moses as the ultimate Proper Authority.

Anyone who has not recently read the words of Exodus 20 might freshly be struck by how deadly serious these first two Commandments seem to be. I paraphrase: If you worship other gods, you are showing hatred for the true God, rather than the required love, and for such a failure you can expect a terrible if unnamed punishment to be painfully inflicted upon yourself and your descendants down to your great-grandchildren and your great-great-grandchildren. One might also be struck by how seemingly arbitrary this announcement is. One is tempted to ask, why stop at punishing the children of the third and fourth generations? What about children of the fifth? And how could a just and fair deity deliberately punish innocent children for the sins of their adult progenitors? These impertinent questions I will leave for the theologians to answer; in any case, the main thrust of the text in Exodus is not to threaten the people of Israel. Rather it is to convince them that the god being introduced by Moses is (unlike the golden calf introduced by Aaron) the real thing: the True God and therefore the Appropriate Object of Their Obedience.

Yes, the persona of the god introduced here is raging and dangerous. He is, as he tells Moses, a "jealous" god, and that jealousy makes him frightening. But those are just some of the negative signs of his earth-shaking, supernatural power; and that power is, or can be for those who choose to obey his authority, ultimately and essentially a positive power. God has already shown to the Israelites that he has the power to free them from pharaoh's oppressive rule. And if (so God now declares to Moses) the Israelites will respect and obey him—will follow his rules and laws and various instructions for worship—he will continue using this unparalleled power on their behalf. God promises that he will "cast out nations before you, and enlarge your borders" (Exodus 34:24). He will drive away the Amorites, Canaanites, Hittites, Perizzites, Hivites, and Jebusites. And if the people of Israel will continue to respect his authority and obey his commands, he will demonstrate the coherence of this positive power by maintaining this wonderful shower of benefits indefinitely. God is arguing for an intimate and long-term personal relationship between himself and the Israelites.

A second look at the closing portion of the second Commandment will suggest at least one reason why obedience to the true God seems so essential. The Commandments are self-referential, so that, in a logical sense, they become an infinitely regressive set of instructions. One might summarize the logic thus: You must obey and love the true God, which means keeping the Commandments, which include the instructions that you must obey and love the true God, which means keeping the Commandments. . . . And so on, ad infinitum. This endless loop or self-referential knot becomes all the more coherent once we see that the third and fourth Commandments instruct us to reinforce God's authority by showing profound deference, both in speech ("You shall not take the name of the Lord your God in vain") and weekly routine ("Remember the sabbath day, to keep it holy"). Given the quality of daily language and the style of weekly routine among some contemporary people who identify themselves as Christians, one might imagine that Commandments three and four are not so important. In fact, the penalty for not keeping the Sabbath was death, according to Exodus 35:2.

Why is obedience so important here? Most people understand why intuitively. Since we recognize the Ten Commandments as a logical whole, a knot, as I have called it, we can also see that this knot might come untied if someone successfully challenged the initial premise. It's a horrifying thought. Or at least so it seems to many, many people, who fervently believe that among the most important functions of religion is the creation and maintenance of human morality. Religion does this extraordinarily important thing, we tell ourselves, by clarifying and enforcing the moral codes handed down by God to man at some point in the historical past. In the Judeo-Christian tradition this belief often presumes that the essence of God's moral code has succinctly been summarized in the Ten Commandments. We can perhaps say that the one billion or so people who count themselves members of this tradition regard it as an article of faith that the initial four Commandments—establishing the God of Moses as the real master or Proper Authority—constitute a keystone in the arch of human morality. Take away the keystone, and the arch collapses.

Or does it?

Within the internal logic of the Commandments, the arch should collapse. But in the real world in which we find ourselves, it does not. Oddly enough, while roughly one sixth of the world's human population believes in the biblical version of the Proper Authority as promoted by Moses, the remaining five sixths of humanity still seem to operate normally and sufficiently. Morality and moral behavior still seem to emerge—even among those who believe that the Proper Authority is actually Allah. Or Buddha. Even for those who hold a more abstract version of the Proper Authority as the Tao. Even for those who see the identity of the Proper Authority as a swirling collective of Hindu or animistic deities.

Who is the Proper Authority reigning supremely and supernaturally above all humankind? What is his (or her) name (or names)? This is an identity problem, and one can only be amazed by how seriously people take it. People devote their lives to spreading their own beliefs about the identity of the Proper Authority. People argue, threaten, fight, torture, and sometimes kill each other over the matter. In any case, the true identity of the ultimate Proper Authority is a theological problem, and I'm concerned here with a very different kind of problem, which has to do with the act of submission itself. Instead of being a theological matter, this is really a psychological one. Let me express it in the form of a question: What inclination embedded in the human mind leads such a large majority of people, people everywhere in the world from every culture and tradition, to pledge obedience to a Proper Authority of the supernatural kind? You probably noticed that I said "large majority of people" rather than "all," because atheists are also among us, people who blissfully reject the notion of any invisible, humanlike deity or deities operating behind the pasteboard mask of material reality. But once we begin talking about submission to proper authority in psychological rather than theological terms, we also see that the inclination has a deeper reach. The role of a proper authority or authorities in human affairs reaches far beyond and beneath that of any supernatural deity or deities.

For starters, the authority embodied in any one of the hundreds of gods revered by humans almost automatically extends to the persons publicly extolling that authority. Take Moses, for example. By showing himself to be the chief interpreter of the ways of a powerful god within the invisible, supernatural realm, Moses is simultaneously showing himself to be an important authority within the visible, natural realm of human affairs among the Israelites.

This aspect of the psychological principle I might summarize thus: The authority to whom people will willingly submit themselves need not be invisible or supernatural, a god or a group of gods. The authority can also be an ordinary person. Such a person might claim extraordinary contact with the deepest powers of the universe—but not necessarily. Historically, many millions, even billions, of people have submitted to the authority of Marx and Mao. And billions more, including those who believe in some kind of deity and those who do not, still submit to myriad other forms of seemingly proper authority on Earth, including the authority represented by one's own parents, by kings or emperors or democratically elected presidents, and so on. The human world is absolutely filled with authorities, and we find that human social systems are given coherence by a vast series of often interlocking authority systems, invisible lines of power running between priests and kings, from parliaments down to principals and parents.

But perhaps you object—insisting that there is an unutterably vast difference between the authority of the all-powerful God and Creator of All Things and that of a mere person, even if the person happens to be a pope or a high priest or imam or rabbi, or a king or an emperor whose own power is buttressed by claims to embody the will of God. But at a psychological level, our feelings about authority are actually interconnected or continuous. Whether it's the authority of God or a parent, a king or priest or judge, our submission to that authority remains generally constant, similar enough from one instance to the next that even in Moses' account we find the distinctions tend to blur and blend. The first four Commandments, all reinforcing obedience to Moses' God, are followed by that fifth Commandment, requiring a similar obedience to parents. And indeed, throughout the biblical text, God is routinely described as Father or Lord or King—as if those three human roles are close symbols or types of the unseen, overarching authority embodied by the deity himself.

If we, for a moment, turn off the sound track, so that we are no longer listening to the static-filled chaos of human chatter but instead looking strictly at the stately dance of human behavior, we will find certain clues that show how universal the psychological inclination of obedience to proper authority is. Those behavioral clues tend to be of two sorts: alignment and posture. Whether we are watching people worshipping the Christian God in a church or the Jewish one in a synagogue, worshipping Allah in a mosque or Buddha in a temple, we are watching people who first align themselves in the direction of power or its symbols, then shift into a posture universally recognized as expressing submission. Those alignments and postures are identical or nearly identical to the alignments and postures of people expressing obedience to the authority of a powerful king, emperor, dictator, or some other sort of potent autocrat. In modern democracies, where power is more broadly distributed and depersonalized, the behavioral responses to power are concomitantly more subtle, so that instead of facing and bowing to the person in charge, for example, people face and take their hats off before a president or senator or judge—or, more symbolically—face and place their hands on their heart before the majestically shimmering, fluttering flag that represents the collective power and glory of their nation.

How seriously should we take these various actions and potential actions? How important is obedience? Perhaps we can best measure its importance by considering the consequences of disobedience. Try standing when everyone else is bowing down. Try facing in the opposite direction from all the other worshippers in a church or synagogue or mosque. In a theocracy, your rebellion may or may not be tolerated, but it will surely be regarded as a serious challenge. In a secular democracy, where religious dissent is accepted by long-standing tradition, you may merely be regarded as a rude or misguided person, ill-mannered rather than immoral. But burn the flag or the constitution and see how many people feel you've done something seriously wrong. Try defying the authority of the state or its human representatives in some meaningful way, and you could find yourself reviled as a criminal or traitor and, conceivably, jailed or executed for your transgressions.

CHAPTER 6
Violence

> But no doubt the first man that ever murdered an ox was regarded as a murderer; perhaps he was hung; and if he had been put on his trial by oxen, he certainly would have been. . . . Go to the meat-market of a Saturday night and see the crowds of live bipeds staring up at the long rows of dead quadrupeds. Does not that sight take a tooth out of the cannibal's jaw? Cannibals? who is not a cannibal?
>
> —Herman Melville, *Moby-Dick*[1]

It was a family enterprise. A wayward mother introduced her daughter and son to the enticements of a criminal lifestyle, and within a short while the entire family was engaged in sinister acts of violence that would terrorize the neighborhood for years. It was the case of Passion and her daughter and son, Pom and Prof: a family of killers and cannibals.

Jane Goodall first became aware of the Passion family's disturbing proclivities in August of 1975. While at her home in town, she heard the news in Swahili over the two-way radio from one of the field staff at the research camp: *Chimpanzee Passion has killed and eaten the baby of chimpanzee Gilka.*

Before that moment, Passion had seemed to be an ordinary female, a mother of average size and middling social rank, who had never done anything particularly unusual. But as Goodall learned, Passion had snatched away the three-week-old baby of another community mother, Gilka, killed Gilka's baby with a bite to the skull, and consumed the flesh and even shared portions of it with her daughter, Pom, her son, Prof, and a young orphan named Skosha. It was bizarre . . . and only the beginning. Gilka soon became pregnant, and about a year later, in the summer of 1976, she delivered a son who was named Orion. In early October, Passion leaped on Gilka, while Passion's adolescent daughter, Pom, grabbed and ran off with little Orion, killing him with a single bite to the forehead. With Passion then taking possession of the corpse, the whole family, once again accompanied by orphan Skosha, fed for the next five hours, biting, tearing off pieces, chewing slowly.

We'll never know what the other members of the community thought about this renegade family and their ugly deeds, but it's possible that some of them found the cannibalism threatening or disgusting. In an October 4, 1976, personal letter to her family in England, Goodall described the reaction of female Sparrow and her daughter, Sandi, to the Passion family meal: "Sparrow . . . came along, picked up a bit of meat, after staring and staring, sniffed it, flung it down and vigorously wiped her fingers on the tree trunk. Her daughter, Sandi, did exactly the same."[2]

Gilka was not the only mother victimized by the Passion family. About a year earlier, another community female, Melissa, had lost a baby under what were then mysterious circumstances. One day, members of the scientific field staff were alerted by the sounds of a raging conflict some distance away from the research camp. When they reached the scene, they found the body of Melissa's baby, killed by a bite to the head—modus operandi, it later became clear, for Passion and family. The field staff found Melissa huddled in the middle of a party of adult males from the community, as if seeking protection, seemingly disturbed by the presence of Passion and Pom lurking darkly nearby. Had the big males intervened during an attack by the two cannibalistic females, only not soon enough to save Melissa's baby?

Melissa became pregnant once again, giving birth to another baby in October 1976. By November, Passion and Pom had killed and eaten that baby as well.

In October or perhaps very early November 1977, Melissa gave birth once more, this time to twins. The rarity of twins among chimpanzees may be related to the cumbersome challenge they create for the mother. It's hard to feed and care for both. Then there is the occasional need to flee from predators—in this case, Passion and family. In a November 8 personal letter to her family in England, Goodall described the attempt at baby-snatching made by Passion and Pom. Melissa had been sitting in her tree nest with a newly expanded family: the twins (named Gyre and Gimble) and her seven-year-old daughter, Gremlin. Gremlin was gently grooming her tired mother when suddenly the killers arrived.

Gremlin's face curled into a grin of fear, and she alerted her mother with a touch. Melissa sat upright in the nest, stared down, emitted a squeak, and dropped quickly down the tree, meeting Passion's son, Prof, at the bottom. She immediately pushed him down and bit him on the neck. Pom approached next, but when Melissa threatened Pom, she hardly reacted. At that point, the field assistants noticed that Pom had her own problems. For some reason, as Goodall wrote in the November 8 letter, Pom was "COVERED in wounds. Face, ears, head, arms, legs, hands, palm, fingers, feet, soles, toes. Only her chest and tummy were unmarked. She was lame, and it gave her much trouble. What a fantastically lucky thing for Melissa!"[3] And Passion, meanwhile, was by then hugely pregnant, possibly too far gone in that direction, Goodall thought, to bother with fighting over someone else's baby.

Passion gave birth to her own baby near the end of the year. Goodall, hoping that a dose of maternal inspiration or hormones would moderate the older female's violent tendencies and return peace to the community, named Passion's new son Pax. That optimistic symbolism was dashed in June of 1978, when Passion made a serious attempt to grab baby Mo from mother Miff. Miff, however, was able to drive Passion away, while a nearby Pom was apparently frightened off by the intervention of field assistants.

So motherhood did not end Passion's evil ways, but would grandmotherhood? Around the end of 1978, her daughter Pom gave birth to a male who was soon named Pan. One day the new mother, cuddling her sweet little offspring, looked up to see her very own baby-eating mother approach—Pan's grandmother—and Pom, quite sensibly, fled with the baby. She returned an hour later and soon relaxed in the presence of Passion, but it took some time, as Goodall wrote home, before Pom would allow her brother Prof "anywhere near her precious baby." Goodall continued: "I wonder—maybe this will be the end of the infant killing for a while. I hope so!"[4]

Infanticide! Cannibalism! Violence at this extreme appears to be the very antithesis of morality, and we might refer to the sixth commandment (Exodus 20:13), "You shall not kill," as our guiding principle. Killing is bad. Killing is wrong. Killing is immoral.

The one realm of social contact where we humans are most convinced that our morality elevates us from the rest of the animal kingdom is the realm of violence. We usually see human morality—whatever its origins—as having introduced a remarkable degree of cooperation, understanding, and nonviolent peacefulness in our own species, whereas we typically discern among animals no such moderating influence whatever. Even those who think in evolutionary terms still like to imagine an exemptional distinction between us and them in the realm of violence.

Anthropologist Sarah Hrdy's superb book on the evolution of human cooperation based on "mutual understanding," *Mothers and Others* (2009), opens with the extended image of an ordinary commercial airlines flight: two or three hundred individuals moving in an orderly fashion into the interior of a large metal tube and tolerating their bumpy passage, under cramped conditions and among complete strangers, to a new destination. As Hrdy reminds us, such an experience can be seriously unpleasant, with strangers bumping against each other, babies howling grievously, and seat neighbors gratuitously pressing flesh against flesh. In spite of all the discomforts and stresses, however, restraint, friendliness, and compassion prevail; and at the end of it all, the same two or three hundred individuals pass in an orderly fashion out of the tube and back onto terra firma. In evoking this scenario, Hrdy promotes her thesis that we humans are a hypersocial species with high degrees of self-control, other-awareness, and empathy. These are evolution's psychological gifts that, she declares, decisively distinguish us as being markedly less violent than the other primates, including our nearest relatives, the great apes.

> Descriptions of missing digits, ripped ears, and the occasional castration are scattered throughout the field accounts of langur and red colobus monkeys, of Madagascar lemurs, and of our own close relatives among the Great Apes. Even among famously peaceful bonobos, a type of chimpanzee so rare and difficult to access in the wild that most observations come from zoos, veterinarians sometimes have to be called in following altercations to stitch back on a scrotum or penis. This is not to say that humans don't display similar propensities toward jealousy, indignation, rage, xenophobia, or homicidal violence. But compared to our nearest ape relations, humans are more adept at forestalling outright mayhem. . . . With 1.6 billion airline passengers annually compressed and manhandled, no dismemberments have been reported yet.[5]

Hrdy's argument looks compelling. Her claims seem reasonable. Unfortunately, however, she simply fails in this example to identify the elaborate edifice of architectural, legal, logistical, political, and technological systems we humans routinely use to reinforce evolution's imperfectly reliable gifts. Think about it: Before every commercial airlines flight, passengers are officially computerized, identified, interrogated, numbered, checked, x-rayed, metal-detected, searched, rechecked, organized, checked a third time, then buckled into an assigned seat and instructed not to move during the most critical times of the flight. You may be among strangers, but there is little genuine anonymity and certainly nowhere to hide. If you spend too long in the toilet, people will notice. If you seem unduly anxious, people will start to wonder. And if you say the wrong thing, make the wrong joke, or violate any of several standard instructions, the gentle smiles and soft authority of the flight crew can quickly be replaced by the harsh manners, handcuffs, and guns of the local and federal police. *Have a nice flight!* is our usual invocation, but for most people, taking an ordinary commerical airlines flight is the closest they will ever come to living in an extreme police state.

So, yes, we may have succeeded in making plane flights civilized experiences with plenty of smiling cooperation and little to no violence most of the time, but we do so at an enormous cost. For all our high intelligence, our adherence to moral codes, our hypersocial and empathetic natures, we *Homo sapiens* still struggle with the same dynamic conflict that animals also struggle with: the one between cooperation and a competition that can boil over into ferocious and sometimes catastrophic violence.

The ever-present possibility of violence among individuals living in social groups has, for humans and for many other species, led to the evolution of psychosocial rules that work to inhibit or channel it. Because they express the social perspective, these rules often inhibit antisocial violence, yet in some instances they may also channel and even encourage violence that is pro-social—violence that appears to serve the community. That violence comes in two varieties, antisocial and pro-social, may seem to you a rather odd idea, since we're all used to thinking of violence as singularly bad and immoral, but in this chapter I intend to explore that concept—the duality of violence—more fully.

Antisocial Violence. Passion's career as a killer and a cannibal provides a fair instance of antisocial violence, and I think it's reasonable to ask: Was she a chimp version of the human psychopath or criminal? At the least, we can say that her acts were unusual. Just before she began her quiet killing spree, a three-year study of the predatory habits of the Gombe chimpanzees in two communities found that the adult females rarely hunted. Around fifteen adult males were actively hunting, though, and they killed an average of 205 prey animals per year. The prey species ranged from rodents to juvenile bushpigs and bushbucks to monkeys of four different sorts, with red colobus monkeys accounting for about three quarters of the total kills.[6] Passion's killing of one to two chimp infants annually, then, would amount to 1 percent at most of the two communities' total predatory activity—assuming that "predatory activity" is the best way to describe her actions. That she was female, while males are usually the active predators among chimpanzees, makes her peculiar acts doubly so.

As later events made clear, Passion was not quite so odd as she at first seemed. During the decades following the first reports of this female's violent inclinations, researchers have documented some predictable patterns of infanticide (followed in some instances by cannibalism) among a significant number of mammals, including several primate species.[7] Among chimpanzees, we now know, infanticide is rare but not anomalous, and it can be perpetrated by both males and females operating as individuals or in small groups. The total number of documented deaths from infanticide among chimpanzees now amounts to around three dozen, a figure based on scientists' observations in four different chimp communities from three different geographic locations in East Africa.[8] It now appears that the logic of chimpanzee infanticide, followed sometimes by cannibalism, is a complex one, probably involving more than one sort of competition that has turned violent.

Such acts and facts may be disturbing, but they do not demonstrate a chimpanzee world devoid of restraint. As the example of our own species demonstrates perfectly well, the presence of violence is not evidence of the absence of morality. We are our own favorite example of a moral species, yet we are also a species capable of chilling acts of violence. If we see our own morality serving as a barrier against violence, therefore, we must admit that the barrier is decidedly imperfect.

That animals also have barriers (likewise imperfect) against killing members of their own kind is not a new idea. In his 1963 book *On Aggression*, the Nobel Prize–winning ethologist Konrad Lorenz introduced the idea that social animals possess "mechanisms" designed to inhibit lethal aggression against conspecifics. When males compete against other males for access to fertile females, Lorenz declared, animals can be explosively aggressive toward each other, but evolution has designed various inhibitory systems that ordinarily keep such aggression from turning lethal. Lorenz described such inhibitory systems as "behavioral analogies to morality," and he argued that those species combining high levels of sociability with strong predatory inclinations and good weaponry should be the ones with the greatest inhibitions of all. Thus, one finds the "strangely moving paradox" that "the most blood-thirsty predators, particularly the Wolf . . . are among the animals with the most reliable killing inhibitions in the world."[9]

Lorenz naively believed that evolution would automatically promote the welfare of the group, a concept many evolutionary biologists now believe to be mistaken, and he was writing well before the full extent of intraspecies violence had been recognized. But his underlying logic—that sociality and antisocial violence are an evolutionary unstable combination—still makes sense. Lorenz took his examples mainly from cases of male-against-male competition, but a cleaner example might be found in the concept of predatory cannibalism.

We can think of *predatory cannibalism* as identical to any other predatory activity, except that the prey coincidentally belongs to the same species as the predator. Predatory cannibalism thus gives us the hypothetical example of antisocial violence without restraint. It's the sort of behavior you would logically expect to find in a predatory animal lacking any moral system.

Cannibalism, carried out through various means and to various degrees, is not uncommon in the zoological kingdom. But much of that cannibalism (often infanticidal cannibalism) is mainly an act of competition—done, in other words, not as a random response to hunger, but more as a targeted attempt to gain competitive advantage: to discourage, displace, or devastate a competitor. Sometimes for some species, infanticidal cannibalism is the consequence of extreme female-against-female competition, perhaps over a good feeding area. Other times, it may be the result of an extreme competition between males, where an upstart male moves to drive out and displace a resident male who has fathered young. And sometimes, an act of cannibalism may follow an episode of violence between mutually antagonistic communities, as in the warfare of chimpanzees . . . or, historically, of humans.[10]

Among the nonhuman primates, eleven species have some record of cannibalism.[11] Unsurprisingly, these species are also, for the most part, omnivores, not vegetarians. Still, another twenty primate species known to be omnivores with significant meat in their diets have no record of cannibalism, so we might conclude that at least these species possess some kind of inhibition against eating their own kind. And what about the human primate? Do we also have innate inhibitions against cannibalism?

Historically, cannibalism among humans may have been widespread. The evidence is convincing enough, at least, for anthropologist Marvin Harris to conclude in his book *Our Kind* (1989) that we do not possess "any natural aversion to the consumption of human flesh."[12] Harris describes the best-known case of routine human cannibalism, that of the Aztecs, whose cruel gods were said to crave human meat and whose high priests therefore specialized in human sacrifice followed by ceremonially distributed consumption. Mainly, the victims were captured in wars with neighboring societies. They were brought back to the Aztec capital of Tenochtitlán, marched up the steps to the high, flat top of a pyramid, and spread-eagled onto a stone altar by four priests, one for each limb, whereupon a fifth priest would slice open the chest cavity with an obsidian knife and surgically excise the victim's still-beating heart. The heartless body was then rolled down the pyramid steps, decapitated, butchered, and—during an elaborate ceremony complete with dancing and communal feasting—eaten.[13]

Cannibalism among the Aztecs consumed many tens of thousands of victims, possibly hundreds of thousands, and Harris argues that the Aztecs were driven to it by economic circumstance: a fundamental shortage of meat. Their more ordinary sources of wild and domestic animal protein were insufficient, he insists, to sustain good nutrition for the one and a half million people living within twenty miles of Tenochtitlán. According to Harris's thinking, then, the Aztecs practiced cannibalism for nutritional reasons, rather than, say, as a dramatic expression of bellicosity done merely to terrify enemies—and you might imagine it resembles the hypothetical case of entirely unrestrained, predatory cannibalism. However, even the cannibalism of the Aztecs must have been powerfully limited by some kind of fundamental psychosocial barrier. After all, only priests were allowed to make the kills, and then only in public and under the guise of placating the greatest powers of the cosmos. The victims were selected carefully and apparently consisted only of social outsiders, not insiders. They were individuals from other communities captured in war or, if from the same community, taken from the disadvantaged classes. Nothing in the story of Aztec cannibalism suggests that it was carried out with the casual, practical, methodical approach we find in simple predation: the opportunistic hunting and killing of animals for food. Even the

Aztecs must have recognized a tremendous psychological difference between killing and eating a person, and killing and eating a deer.

Actually, it's hard to imagine pure instances of predatory cannibalism—cannibalism that's psychologically indistinguishable from predation—among social animals in general, because the concept evokes the following contradictory situation: a coherent society consisting of incoherently antisocial individuals. At best, we are describing an evolutionarily unstable combination. If all chimpanzees made no distinction between chimp and nonchimp as a source of meat, what chimp would choose to join a group, submit to grooming, or fall asleep in the presence of others? And under such circumstances, how long could any chimp society continue to exist? Cannibalism without restraint will inevitably weaken or break social bonds. If it exists at all, therefore, it should be found only among generally asocial species.

Take spiders, for example. For numerous spiders and other invertebrates, the social act is limited to the sexual one—which, for many species, cannibalism endows with a distinctive frisson. For a few, the cannibalism happens before fertilization. That might seem to be an oddly self-defeating activity with no evolutionary logic whatever. In fact, the logic is probably female-centered. It could be an emphatic way for her to say no, an instance of female choice that includes nutritional benefits. She can improve her fecundity while saving her eggs for a better male.[14] For many cannibalistic spiders and other invertebrates, though, the female prefers to dine after sex, or, as in the case of several species of predatory ceratopogonid biting midges, during. In this instance, once the eager couple joins in a nuptial embrace, the female penetrates the male's protective exterior and proceeds to liquidate his nutritious interior. For her, cannibalism provides nutrition, while for him (intensely competing with swarms of other eager males), it's an effective way of keeping her undivided attention for the half hour it takes to fertilize all her eggs.[15]

Gastropods—approximately sixty-seven thousand species of slugs and snails specializing in marine, freshwater, and dry-land environments—include at least a couple hundred predatory varieties that are known to be cannibalistic, some indiscriminately so. In some species, the nutritional benefits of cannibalism must be compelling indeed. For one type of Swedish land snail, for instance, eating a conspecific's egg provides enough nutrition to enlarge the eater's own shell by one quarter in three days, a rate of growth that ordinarily takes three weeks, and it increases the eater's probability of reaching adulthood by around 40 percent.[16] So it's clear why some or many predatory species of gastropods will turn to cannibalism (of full adults or their eggs) as an easy and important source of nutrition. The startling fact is that many do not. A number of predatory gastropods rarely or never consume conspecifics, which suggests that even the lowly slugs and snails of this world can recognize members of their own kind. It also raises the possibility that they are following some system of inhibition or restraint that keeps them from opportunistically killing and eating one another.[17]

Mammals have the cognitive equipment to make quick and fine distinctions between species, and most have social inclinations that in ordinary circumstances ought to inhibit cannibalistic ones. Still, you might be thinking that the lack (or minimal presence) of predatory cannibalism among chimpanzees and many other mammals who are both predatory and social—such as lions and hyenas—merely reflects situation and opportunity. Predators prefer prey who can't fight back. This simple, pragmatic fact, you might insist, will explain all there is to know about a predator's choice of prey and the reluctance of individuals to prey on conspecifics. But chimpanzees, lions, and hyenas are all patently capable of acting collectively, in groups of two or three or more, to kill lone adults of their own species; and they routinely find themselves in a social community among other adults and juveniles who, asleep or relaxing or distracted or weakened by disease, are temporarily vulnerable to a quick and easy kill. Yet these animals seldom take advantage of such vulnerability within their community, and they almost never do, or so it would seem, purely for the sake of fresh meat.

The psychological systems inhibiting cannibalism in social species could include an evolved distaste for killing one's own kind, as well as an evolved disgust at the appearance or taste or smell of conspecific meat. They will probably include an evolved psychological reluctance to harm individuals who could turn out to be genetic relatives—likely enough in small societies—and also the evolved reluctance to harm friends or allies. Additionally, we can say that many of the supposedly pragmatic situations inhibiting cannibalistic attacks are not actually physical circumstances or randomly occurring social accidents at all but are, instead, themselves the products of an evolved psychosocial environment. One such inhibiting situation: mothers primed by evolution to protect their offspring. Another inhibiting situation: relatives and friends and allies also primed by evolution to protect their own from antisocial violence. Furthermore, unrestrained cannibalism can bring a significant health risk. The closer the eater is to the eaten, in a phylogenetic sense, the more likely that a disease-causing organism, fully adapted to live in the eaten species, will find a welcoming home in the eating one.[18] Such a health risk could also produce evolutionary effects inhibiting routine cannibalism.[19]

Was Passion a predatory cannibal, a crafty female limited in her feeding activities solely by physical circumstance and random opportunity, routinely ready to consume anyone indiscriminately—chimp or nonchimp, friend or foe, kin or nonkin, infant or adult—as soon as she had a decent chance? It's possible. But if so, she would have been the chimpanzee version of an antisocial psychopath or criminal. Her story still evokes a powerful contrast to that of the rest of her community, including the most active predators, the big males, who during this period never followed her example.

Pro-Social Violence. Primatologists Robert Sussman and Paul Garber, collaborating with geneticist Jim Cheverud, analyzed the data on social interactions among sixty different primate species, including apes, monkeys, and prosimians (such as lemurs and bush babies), and concluded that these animals spend about 10 percent of their daily activities in social interaction. More than nine out of ten of their social interactions, moreover, are positive, acts promoting affiliation (friendliness) and cooperation, while fewer than one out of ten of their social interactions are overtly competitive or aggressive.[20] Affiliative and cooperative behaviors among apes are typically grooming and play. Overtly competitive or aggressive behaviors include anything from "mild spats, displacements, stares, and avoidance" to "chasing, fighting, and biting which can result in severe injury, death, and social disruption."[21] Sussman et al. found such hostile interactions among prosimians to occur at an average rate of 0.16 events per hour. Among monkeys, they happened at an average rate of 0.6 times an hour, while among the apes, such events happened at the "extremely low" rate of only about 0.09 times per hour. Roughly, then, an individual ape experiences little more than one unpleasantly competitive or aggressive event every waking day.

For anyone imagining that the world of animals amounts to one violent scramble for resources—a tooth-and-nail, dog-eat-dog, every-animal-for-himself sort of place—these figures may be startling and should be profoundly educational. Unfortunately, though, the study vastly underrates the destructive effects of even rare acts of violence (how many friendly grooming episodes will balance a single killing attack?)—and it relies on the unexamined assumption that cooperation and aggression are invariably opposites. They are not.[22] Animals of many kinds can often be cooperatively aggressive and sometimes cooperatively violent.

Take lions, for instance. Lions would seem to be the poster kittens of animal cooperation. Their social lives center on prides, groups of several related adult females accompanied typically by a small coalition of adult males who are unrelated to the females. Prides in Tanzania's Serengeti National Park have been known to include as many as eighteen adult females and as few as one, so there is a tremendous variety in pride size. The males are socially peripheral; the females form the critical social core of the pride; and any females who are nursing cubs gather into communal maternal groups called *crèches*. You might say they mother cooperatively. It's clear that they hunt cooperatively, in teams. Lions even sleep cooperatively—or at least communally, in big furry piles. None of this is surprising, since the gregariousness of lions in a pride can easily be observed through a weak set of binoculars. The greater details of their cooperation have been teased out only during the last few decades, through intensive scientific scrutiny and analysis. We now know that lions, at least the females, are not merely effective cooperative hunters but also strikingly egalitarian ones. No adult female is dominant over any other, and when they're feeding on a freshly killed carcass, all will have equal access to the meat, no matter how much or how little each participated in the kill.[23]

It's an appealing instance of motherhood and sisterhood at work in an animal society: cooperation in the service of the community. But cooperation for what purpose? The traditional notion has been that cooperative hunting among lions makes nutritional sense because a larger group of hunters can bring down more and bigger prey than a smaller group or a single individual. Yet, the larger group has to divide the meat more ways than the smaller group or the solitary hunter. According to a study organized by biologist Craig Packer, who has directed the Serengeti Lion Project for the last thirty-one years, a solitary hunter with only herself to feed is nutritionally quite efficient. Indeed, according to Packer's nutritional analysis, solitary hunting should be relatively common among lions—but it isn't. Why? One clue: Solitary females have a distinctly higher rate of mortality.[24]

Lion prides maintain territories, and they do so through competition with neighboring prides, in which group numbers are important, so that a lone female or a small hunting team from one pride always has the risk of a fatal encounter with a larger team from a neighboring pride. Packer's research finds that aggressive and potentially violent interpride encounters occur an average of once every 116 hours, though mostly they end with chases rather than actual fights.

The crèches likewise don't make sense nutritionally. The mothers in a crèche tend to hunt together, and they don't (as, for instance, elephants do) share babysitting or nursing. Indeed, the grouping in the crèche seems to offer no nutritional advantage at all to the cubs, and in the larger crèches—up to nine mothers and their cubs—the cubs may appear to be undernourished. Yes, crèches are communal endeavors, but, as with the hunting teams, these groups appear to exist mainly for defense against the violence of other lions. *Other lions* means males from beyond the community who form their own cooperative groups and search for a pride with defenses weak enough that they can take it over. Taking over a pride will require cooperative violence on the part of the outsider males, who will fight against the violently cooperating insider males, the ones guarding the pride; should the intruders kill or drive off the pride males, the intruder males, each twice the size of any adult female, will then descend on the crèche. The crèche females will cooperatively fight to drive away the outsiders, but if they fail, the invading males will then, routinely and with cold intent, kill all the youngest cubs and chase away the older ones and the adolescents. This is infanticide, carried out by invading males from another social group, and you might imagine it to be rare or anomalous, but it's not. It's part of an established pattern that accounts for 27 percent of all infant mortality among the Serengeti lions.[25] That is significant violence, perpetrated cooperatively and resisted cooperatively.

This predictable infanticide is not done out of spite or as a result of some random brutal rage. Rather, it expresses a clear evolutionary logic, since soon after the adult females lose their cubs, they will cycle back into fertility and become ready to mate with the new pride males, the successful invaders who have proven themselves in the most direct way possible to be more powerful and better warriors than the defeated males—and thus more fit as mates. In short, the entire social structure of a Serengeti lion community is shaped, through evolution, by the powers of cooperation and by the periodic fact and continuous threat of violence from other lions: what biologist Packer calls "the dreadful enemy."[26]

Spotted hyenas are another cooperatively violent species. They live in what researchers call *clans*, large communities that can include dozens of individuals and are led by females. Spotted hyenas are matriarchal, in other words, and the sisterhood of the clans is expressed in cooperative mothering and a daily life filled with warm greetings, frolicsome play, mutual grooming, and opportunistic sunbathing sprawls. At night they hunt in groups that show a sophisticated and intelligent sort of cooperation. Spotted hyena clans are run by the females in part, probably, because the females are tightly bonded genetic relatives while the males are unrelated transfers from other communities. Physically, the females are moderately larger than the males, and they have about the same high levels of testosterone as the males. In good hyena country, each clan holds territory that will be contiguous with the territory of a neighboring clan, and in spite of the obvious affiliation and cooperation that inform life *within* the clan community, life *outside* that community—relationships between clans—is invariably competitive, often aggressive, and sometimes violent enough to produce killing raids of neighbor against neighbor. These attacks can be predicted in situations of tactical advantage, when a larger group of hyenas from one clan meets a smaller group or a lone individual from a neighboring clan.

Hans Kruuk, the scientist who first reported territorial battles between rival spotted hyena clans, has described one encounter that occurred in 1967 at his study site in the Ngorongoro Crater of Tanzania, after members of the Mungi clan took down a wildebeest just inside the territorial boundaries of the Scratching Rocks clan. The ensuing melee, a quick clash between around twenty Mungi hyenas and some forty members of the Scratching Rocks clan, led to the strategic retreat of the outnumbered Mungi hyenas and the fatal mauling of a single Mungi male who failed to escape. In Kruuk's words, "The Mungi male was literally pulled apart, and when I later studied the injuries more closely, it appears that his ears were bitten off and so were his feet and testicles, he was paralyzed by a spinal injury, had large gashes in the hind legs and belly, and subcutaneous hemorrhages all over." When Kruuk returned to the site the next morning, he discovered one Scratching Rocks hyena still gnawing on the carcass, while a closer examination showed "evidence that more had been there; about one-third of the internal organs and muscles had been eaten. Cannibals!"[27]

In the world of spotted hyenas, then, trespassing can be fatal, because these intelligent and powerful animals are sometimes, under some circumstances, cooperatively violent.[28]

The social structure of chimpanzees is not so different from that of spotted hyenas, except that chimpanzee communities are male-dominated, rather than matriarchal. But life in a chimpanzee community can be as pleasant as life inside the hyena clan: a coherent group of individuals associating effectively with one another. Chimpanzee communities also have territories defined by invisible borders (or narrow overlap zones between territories), and since a community is always in danger from attack by the neighbors, chimpanzees routinely patrol those borders. Patrols are carried out by males, perhaps a half dozen or more adults, sometimes accompanied by an adolescent male or a female. Patrols can begin with plenty of raucous noise and other signs of excitement, as the members gather and acknowledge each other's supporting presence. At first it may seem like a festive occasion, a party on the move. The seriousness of what they're doing becomes apparent once the patrol reaches their territorial edge and the hairy apes stop still to gaze for many minutes, intensely and in total silence, across to the dangerous foreign lands occupied by a hated neighbor.

The patrol that paused, completely still and silent at the border, has now begun, quietly and cautiously, to proceed into enemy territory. It's a dangerous move with potentially deadly consequences, and although the raiding party cannot count on superior weapons to give them an advantage over the enemy, they do have an underlying strategy. The raiders are looking for signs of the enemy, and if they find him, or them—a rustle in the underbrush, the sight of a face, a hairy arm or leg in the thickets—they will consider their numerical strength compared to that of the enemy. If they have found a lone member of the neighboring community, or perhaps only two, with one of them young or very old, they will probably attack. If, however, the raiders discover an enemy who begins to approach their own strength in numbers, they will flee.[29]

Assuming the raiders have found a vulnerably isolated member of the neighboring community, though, they rush in for the attack. The males have weight-lifters' physiques that now seem to double in size as, in the excitement and rage at seeing a hated enemy, their hair bristles straight out. Thus prepared and enlarged, they race in, corner and trap and grab the offending chimpanzee, pummel him with their fists, kick him with their feet, bash him with a rock, bite, tear, lift and slam him down to the ground. The goal is not to frighten or drive away or punish. The goal is to annihilate.

One could study wild chimpanzees for years and never see clear evidence of war. Unlike lions and hyenas, who live on the open plains and can easily be watched, chimpanzees live in the obscurity of dense forests. To begin seeing chimpanzee life as it really is, you have to reach a point where the smart and perceptive apes stop disappearing every time they see you. That first habituation takes years, and once it happens, you will be introduced to the fascinating, often affectionate lives of chimpanzees within a single community. War, however, takes place at the edges of a community and sometimes well into the territory of a neighboring community, and so to witness it to any significant degree, you have to habituate two communities. That takes time. At this point, the figures on chimpanzee cooperative violence suggest that an individual's yearly risk of death from intercommunity warfare ranges from 0.069 to 0.287 percent. That's generally higher than the risk of death from war for people living in modern nation-states, but it's still within the same general range as the risk of death in war for contemporary people living in pre-state hunter and farmer societies.[30]

We know more about war among the chimpanzees of Jane Goodall's research site in Gombe National Park, Tanzania, not because chimpanzee violence is more common at Goodall's site, but because we know a good deal more of everything about those chimpanzees. Jane Goodall's research project is the longest-running and most fully developed study of chimpanzee behavior in history.

The big war witnessed at Gombe occurred during the 1970s after the main study group split in two, forming two separate and, soon enough, mutually hostile communities on the northern and southern sides of a shared boundary. The eight northern males included six in their prime and two past their prime. The southern group included four males at the height of their physical powers, along with two older males and one adolescent. A number of females lived with the northern community, while three adult females had settled into the southern territory. Since the northern community claimed the Kasekela valley at the center of its territory, they came to be known as the Kasekela chimpanzees, while the southerners were identified by their own Kahama valley.

The first signs of war between the two were recorded on the afternoon of January 7, 1974, when a Gombe field staff member followed eight Kasekela chimpanzees as they moved directly, with seeming deliberation, southward, soon approaching the edge of Kahama territory. That group included Kasekela's top-ranking male, Figan, and his main rival, Humphrey, along with four other adult males, one adolescent, Goblin, and a childless and apparently infertile female named Gigi. Crossing the border and passing now into the outer edges of Kahama-land, the eight Kasekela apes soon discovered a lone twenty-one-year-old Kahama male named Godi, who that afternoon was quietly sitting by himself in a tree and munching away serenely on some fruit.

By the time Godi saw the excited gang from Kasekela careening through the thickets, it was too late. He jumped out of the tree and ran for his life, with three attacking males, racing side by side in close pursuit. They soon caught up, with the big male Humphrey grasping and yanking young Godi down by his leg. Then, crushing him to the ground with his great bulk, Humphrey immobilized his victim, while the other males went to work: biting and pounding with their hands and fists for about ten minutes before Humphrey released Godi, and the attack stopped. A screaming Hugo threw a huge rock at Godi, missing him, and the attackers ran off in terrific excitement, hooting loudly, and leaving their screaming victim with bite marks on his right leg and torso and deep facial wounds. Godi was never seen again. Presumed dead.[31]

The strategy of attacking only with undeniably superior numbers naturally favored Kasekela, which had begun the conflict with more warriors than its neighbor. As gangs from Kasekela picked off the Kahama males, one by one, the numerical balance tipped more and more decisively in their favor. Along with their numerical gains, the Kasekela community was soon benefitting from major territorial gains. In early 1973, the Kahama chimps' homeland consisted of some 10 square kilometers of forest, but by the end of 1974, their home territory had shrunk to only around 3.8 square kilometers, whereas the Kasekelas were soon confidently moving across some 15 square kilometers.

After nearly four years of male patrols and raging gang attacks, from early 1974 to late 1977, the chimpanzee war ended with a complete victory for one community and the complete annihilation of the other. All the Kahama males were killed (or disappeared under violent or suspicious circumstances), as were most of the females, all except for three nubile females who were taken into the victorious Kasekela community. By 1978, with Kahama no more, members of Kasekela, male and female, were traveling and eating and sleeping in an expanded territory that included all of what had once been Kahama lands. Of course, such territorial gains would mean nutritional gains, since Kasekela now had a bigger smorgasbord to choose from. All was well for Kasekela, and if the Kasekela chimpanzees had language and a symbolic culture, we might imagine that the victorious killers would be honored as heroes. They had bravely risked their own lives in support of the community.[32] That's what the chimpanzee history books might say . . . and the same books might also more soberly note that the victorious Kasekeleans now had new neighboring communities and new battles to prepare for.[33]

No one can deny the cooperative violence that periodically punctuates the lives of lions and hyenas, but the story of cooperatively violent chimpanzees has created a minor cottage industry dedicated to denial.

After all, chimpanzees live in forests where they're often hard to see, so conflicting interpretations may be possible. More to the point, chimpanzees are humanity's closest evolutionary relatives, so anything said about those apes might be imagined to reflect, for better or for worse, on humans and human nature. In fact, the picture we have of chimpanzee cooperative violence is well established, but the deniers challenge the data from Goodall's site by complaining that she fed her chimpanzees too many bananas, which, by causing crowded conditions for a few seasons, turned those gentle apes into raging monsters—or, in the slightly more nuanced language of one armchair theorist, somehow "triggered a switch to a negative mind set."[34] The deniers cite in support of their counterarguments data taken from minor, short-term research projects where the habituation of chimpanzees has been nonexistent, incomplete, or limited to only a single community so that the interactions between communities could not be observed. The deniers sometimes resort to sarcastic caricature, and some of them have insisted with grim sanctimony that "most apes" are "remarkably nonaggressive." Given the potential for comparison with humans, they have also insisted that humans living in small, pre-state societies are also "remarkably nonaggressive."[35] Indeed, so these same deniers have announced, anyone arguing for an evolutionary comparison between the violence of humans and that of apes is either naively ignorant or deliberately "misappropriating" anthropology to "reinforce the status quo by positing underlying biological causes to complex social problems."[36]

Such a casual attribution of ugly political motives to those who hold alternative ideas is a signature of the academic ideologue. In this case, the ideology originates in a variety of wishful thinking that would have us see human nature as fundamentally good, with the badness arrived by way of civilization, culture, or various unfortunate social developments. In this version of things, violence has little or nothing to do with the evolution of human (or chimpanzee) nature, and thus such problems as war—being mere social problems with no underlying evolutionary logic—should be possible to fix with the right kind of social engineering. It's a hopeful idea, assuming it's true. But the contrary idea, that human violence also, like human kindness and friendliness and cooperation, contains evolutionary elements, is to be avoided whether true or false because it could lead to fatalistic inaction.

As the deniers write, should the idea that "human problems such as warfare and killing" are "primordial, adaptive and natural" become commonly fixed in the minds of ordinary people, it would diminish "our impetus for positive change."[37] The deniers don't go into details about how they intend to create such an impetus, and they really should, since so far this new century doesn't look any more peaceful than the last. Close to ninety million people died during the twentieth century as a result of war.[38] If we assume that the total number of people who lived during the century was approximately two to three times the number alive at century's end, then we can calculate that one's chances of dying as a result of war hovered at around one tenth of 1 percent. That figure may be taken as evidence that human violence is extremely rare and hence negligible, but you would never know it from the history books or from many people's nightmares.

Innocent bystanders suffer from war, but war begins in the violence of men directed against other men. If we consider the acts of men directed against women—including but not limited to rape, battering, and such broadly applied cultural practices as selective abortion, infanticide, and clitoridectomy—statistics representing the personal cost of human violence will rise markedly. We cannot ignore human violence, nor can we afford to tuck away the reality of human war under the vaguely comforting label of "complex social problems" without some honest further consideration.

We humans use language to comfort and reassure, and we use language to attack and bully. Our ability to talk means that we can make up self-serving stories and comforting fantasies, and it means that we can discuss and analyze the stories and fantasies and even, sometimes, look behind them to find deeper truths. Animals have no equivalent complex symbolic language to talk about what they do and why they do it, so in studying animals, we resemble alien anthropologists from outer space who, coming down to our planet and hoping to understand humans, are unable to crack the code of our language. Not understanding what we say, they are stuck with watching what we do.

Industriously, therefore, they set up a number of observation posts and begin collecting data on human behavior, subjecting it to sound statistical analysis. First, they look at human behavior in general. The alien anthropologists watch fishermen on a fishing trawler with mile-long drag nets haul up catch after catch in the northern Pacific. They observe workers on a Japanese factory ship disassemble endangered whales. They follow an American dentist from Florida as he closes up his office for a vacation and travels into the far north to kill polar bears with a high-powered rifle. They monitor the traffic in elephant tusks from Central Africa to China. The aliens' first accumulation of data quickly clarifies that no nonhuman species on the planet is universally protected from human exploitation.

No surprise to the alien anthropologists! They have themselves emerged on their own planet through evolution by natural selection. They have already harvested and consumed many of the other species there. Maybe they now regret such actions, since their home has become a biologically impoverished planet, but they are intellectually advanced enough to understand the evolutionary principle of Darwinian narcissism. Morality ordinarily extends to one's own species only. So the real question the aliens are curious about is this: How do humans treat each other?

During their first months of study, the alien anthropologists find clear evidence of an attractive sort of biped. They see signs of kindness everywhere. They see people being routinely polite and almost continuously helpful to one another. Yes, the aliens do understand money and the concept of economic activity, and it's clear that a good deal of that apparently nice behavior is actually based on an economy of self-interest. Money changes hands after a service. Nevertheless, the aliens also see indications of kindness and helpfulness when no money is exchanged, and they develop an overriding sense of humans as positive and loving toward one another.

As time passes, however, the anthropologists also begin to notice some less positive things. One of their researchers, stationed discreetly in a dark alley somewhere, witnesses one person killing another deliberately and brutally. This observation is a shock at first, and the initial reaction of the researchers is puzzlement, followed by the idea that the behavior was a mere anomaly with no significance. Maybe it says nothing about the bipedal species. More time passes, more data accumulate, and the alien anthropologists now recognize that rather than being a meaningless anomaly, humans killing other humans happens rarely but still, once they look at a big enough population, regularly. So now the aliens conclude that humans killing other humans is part of the picture. This is still no surprise to the alien anthropologists, though, who quietly note that such antisocial violence happens occasionally among their own kind back home.

But as more time passes, the aliens begin to assemble data that show a peculiar pattern to the human killing behavior. The anthropologists slowly begin to see that humans actually have two radically different categories of killing, and the difference between the two can easily be identified by observing the reactions of other humans who witness such events. The first sort of killing always causes strong, almost universally negative reactions among the people witnessing it. A killing occurs, and people chase, hunt down, and even attack the killer. But a second sort of killing produces completely the opposite sort of response. A killing occurs, and people find, applaud, and even honor the killer. Why this extraordinary difference?

The alien anthropologists are puzzled because this pattern isn't characteristic of their own species, which is psychologically and socially rather like elephants. They do witness the occasional killing among their own kind that seems profoundly antisocial, and in their own language they describe it with words equivalent to "wrong" and "bad" and "evil." But here, the aliens come to recognize, they have discovered a species where individuals not only engage in regular antisocial killings, they also sometimes carry out killings that appear to be cooperative, pro-social acts. How remarkable. How extraordinary. How strange. By this point, the alien anthropologists have done enough research to recognize that the pro-social killings, although they seem to occur in particular places at particular times, are consistent enough to account for almost ninety million human deaths during the last one hundred rotations of the green-and-blue planet around its sun.[39] That degree of statistical specificity marks a breakthrough for the aliens, and it is accompanied by a second one. The anthropologists from outer space have finally cracked the code of human language enough that they now know the word describing circumstances in which such pro-social killings occur. The word is *war.*

During war, the same behaviors that would have landed someone in jail and caused him (and it's usually him, not her) to be branded a callous murderer in peacetime become instead the actions that can reap honor and praise. A good soldier in war is saluted and pinned with medals, whereas the hesitant warrior who decides to refrain from killing is branded a coward or a traitor and can expect to be shunned or even shot. This may seem like an strange reversal of values indeed, with the sixth commandment apparently turned upside down or inside out. Yet few people consider it strange. We typically respect the brave young soldier who has risked being killed in order to kill others on behalf of his community. So the social approval given to killing during war is not merely something cynically arranged by corrupt governments and passively approved by a naive public. Instead, the approval is routinely agreed upon by thoughtful, intelligent citizens and even, with some consistency, by those recognized to be the official interpreters and guardians of our moral codes.

We are used to thinking of war on the grand scale: a nation hundreds of millions strong sending its army, navy, and air force out to meet an enemy who might be equally formidable in size and organization. But it's still war even when only one side has an enormous modern military organization and the other side is represented by a collection of gaunt, grizzled guerillas living in a warren of basements or caves. And it is still war when both sides consist of guerrillas living in basements and caves. So the size of the armies and their degree of organization are not essential aspects of human war either. We can readily imagine a ferociously bloody war waged between two small communities of people who live on opposite sides of a river or a valley and, for some reason, have decided to fight each other to the death.

But isn't morality about being good, about loving thy neighbor as thyself, about helping little old ladies across the street and all the rest? Here we are reminded of a peculiar twist in the morality of humans. The human species includes among its temperamental qualities the powerful tendency to cluster into coherent yet exclusive and sometimes mutually hostile social groups, where each side is capable of seeing the other with a nearly reversed set of feelings and values. This reversal is one of the clearest indicators of war: where one side places its own individual warriors' actions at the very top of the list of positive morality even as the other side sees the same individuals' actions at the very bottom of their list of negative morality. When such feelings and perceptions are expressed in language, we learn that the *heroes* and *freedom fighters* and *martyrs* of one side are actually identified as *killers* and *thugs* and *terrorists* by the other.[40] This is not a simple problem of language and habit, but rather a far more challenging one of emotions and history. Morality tends to focus its values at the level of the community, and the moral psychology of our own species brings to human communities a capacity for high levels of xenophobia, where individuals from one group can fiercely and even reasonably hate and fear the individuals of another. Such is the strange little twist of human morality, strange enough to puzzle the alien anthropologists who have come down from outer space to investigate. It's strange, all right, but it is by no means uniquely human.

CHAPTER 7

Sex

As ashore, the ladies often cause the most terrible duels among their rival admirers; just so with the whales, who sometimes come to deadly battle, and all for love. They fence with their long lower jaws, sometimes locking them together, and so striving for the supremacy like elks that warringly interweave their antlers.

—Herman Melville, *Moby-Dick*[1]

Sex: I had already seen an awful lot of it, and when I spotted that rhythmically swaying clot of torsos and limbs high in the trees, I thought I had stumbled across just one more loving couple. I pulled out the binoculars for a better look. My companion in the forest, anthropologist and chimpanzee expert Richard Wrangham, did the same.

Chimpanzee sex is sometimes brutal, typically businesslike, and invariably brief—but these weren't chimps Wrangham and I gazed at through our trembling tubes. These were bonobos, a separate though closely related ape species; and I was excited to be witnessing an act that appeared just somehow different from the mating of chimpanzees. True, the male had positioned himself behind the female, very much in the chimpanzee style, but nevertheless the whole act still seemed, in some not quite definable way, rather unchimplike. . . . Only after the couple in the trees had finished and separated did I recognize just how unchimplike the behavior actually was. Both participants, I saw then, still had erect penises. So what I had at first imagined to be a male-female coupling high in the trees, a pleasantly lingering heterosexual dalliance, turned out to be two males having a go at homosexual sex. They were too far away for either of us to observe any of the plumbing, but their position and movements suggested anal intercourse, or something close to it.

Sex between females we saw far more often, and it soon began to seem perfectly ordinary. For paired females, sex often takes place in the missionary position. One lies on the ground on her back, legs apart, inviting the other. Her partner moves on top. Face-to-face, eyes open, they embrace. With hips moving from side to side, they rub the sensitive tips of their large clitorises together, slowly at first but with an increasing intensity and frequency until, at last, they clutch and squeal and cry out, apparently reaching a pleasure-filled orgasm.

You can read about bonobos and their remarkable lives in primatologist Takayoshi Kano's book *The Last Ape*, translated into English for a 1992 publication. But since Wrangham and I were just then Kano's guests, having pitched our tents on the ground right next to his little mud-brick house in the Mongandu village of Wamba (north-central Democratic Republic of Congo), we were able to learn about his pioneering study more directly, while sipping from postprandial glasses of Cointreau, in conversations with the author and pioneering researcher himself.

Bonobos used to be called pygmy chimpanzees; they look a lot like chimps while being a little smaller, on average, and more gracile. More distinctive are the body proportions, with chimps having a larger head, thicker neck, broader shoulders, and shorter legs. Bonobos have a more humanoid proportion in the ratio of legs to body, and they also seem to find upright standing and walking more comfortable than chimps do, since they use an upright stance more often and for longer periods, often as a way to free their hands to carry food. Chimpanzees have dark or black hair distributed across their bodies and around their faces, as do bonobos; but bonobos also have an extra mop of hair on top of their heads that is naturally parted in the middle, as if they're all customers of the same tyrannical hairstylist.[2] Chimps are also a lot noisier than bonobos. They call to one another with explosive bursts of hooting, barking, and screaming powerful enough to travel for up to a kilometer in the forest. Bonobo vocalizations are softer, more chattery, and higher pitched,[3] enough so that when Kano first heard them, he thought he was listening to birds "twittering in the distance."[4]

These few differences alone might be enough to mark the division of two species, but bonobos most fully demonstrate their divergence from chimpanzees at the temperamental and behavioral levels.[5] Bonobos have been described as the "make love not war" species, a memorable cliché that fairly summarizes two important and interrelated things about them compared to chimps. First, they are remarkably peaceful. Second, they are very sexual.

Kano's project at Wamba used sugarcane as bait to attract bonobos onto an open field, where their behavior could be studied by scientists. It was a sensible way to do research, but what would happen if individuals from different bonobo communities wanted the same sugarcane treat at the same time? One day near the end of 1986, researcher Gen'ichi Idani watched as bonobos from two separate communities, known as E-group and P-group, arrived at the open field with the sugarcane simultaneously. It must have been a shock for members of these two separate communities to see one another, strangers peering out from the edges of the forest. Slowly, however, the individuals of E-group and P-group emerged from behind their protective barriers of vegetation. Gradually, they came closer. They sat. They stared. They called. It was a tense thirty minutes—until finally a female from P-group crossed the open area between the two groups and had sex with a female from the other side. That seemed to break the ice nicely, and for the next couple of hours the two groups mingled normally and ate the sugarcane in peace.[6]

This marked the start of two months of peaceful meetings between E-group and P-group at the sugarcane site. Invariably, when the two groups moved into that single feeding area, females would initiate the friendly contacts, which included mutual grooming and sex with other females from the opposing community. Females were even seen mating with stranger males, while their own community males quietly looked on. Such behavior is utterly unheard of among chimpanzees, where the males are inclined to kill stranger males and where, within one's own community, who gets to mate with any estrous female is forcibly regulated according to rank within the male hierarchy.

Bonobo males form dominance hierarchies, too, and male-against-male aggression is common enough, but male bonobos are still much less violent with each other than their chimpanzee counterparts. Primatologist Frans de Waal writes in *Bonobo: The Forgotten Ape* (1997) of the day two adult male bonobos, Kevin and Vernon, both living in captivity, were placed into the same zoo enclosure after having been apart for an extended period. Two male chimpanzees in this situation might be expected to descend quickly into a raging, bloody encounter over dominance, and indeed, the two male bonobos began circling each other and screaming.

But Kevin, the younger of the two, extended an arm and began to make a beckoning movement with his hand while rapidly flexing his fingers. Humans make a similar hand gesture to encourage someone to move closer, and a similar kind of finger-flexing is sometimes used for emphasis. Kevin seemed to be asking the older bonobo, Vernon, to move closer. Both of the males had erect penises by then, and they displayed their penises in the style and manner—legs apart, body swaying—commonly used to solicit sex with females, but they remained apart and circling cautiously. "It was as if," de Waal writes, "each male wanted contact but did not know if the other could be trusted." Finally, however, they raced into each other's arms: embracing, grinning, and rubbing their penises together. That was enough. The tension between them evaporated. Their screams turned into eager food calls, and they dropped to the floor of their enclosure and began collecting scattered raisins.[7]

Bonobos illustrate a *Kama Sutra* of erotic possibility for the animal kingdom. Sexual contact ranges from intimate touching and fondling to manipulating one another's genitals to oral sex to tongue kissing. They couple heterosexually, homosexually, in groups; and when no partner is convenient or available, they masturbate.

Heterosexual sex among bonobos begins with a solicitation. Someone signals or proposes, and the other responds. Kano's study suggests that males initiate sex with a quick courtship display about three times more often than the females solicit sex—but then, who knows? Females in a sexually receptive state will often be seen feeding closely to males, who then become interested, so perhaps the female's early action amounts to a first provocation.[8] Anyhow, after various displays, solicitations, and provocations, they mate: sometimes in the front-to-rear style of chimpanzees and sometimes in the front-to-front style that often characterizes human sex. Face-to-face, front-to-front, belly-to-belly sex is called *ventro-ventral* sex by sober scientists.

For the slightly less sober rest of us, it's sex enjoyed in the missionary position, an interesting expression based on the Victorian notion that face-to-face is the most dignified and civilized way of doing such an inherently undignified and uncivilized thing, and that it has the virtue of keeping one's eyes and mind focused on subjects of a more elevated sort. But the missionary position has also seemed to embody a more general virtue. It has served to represent yet one more critical feature distinguishing human from animal. Of course, we now recognize that the missionary position is not a human cultural invention or a mark of human uniqueness, since bonobos also use it. Under natural conditions at the Wamba site, about 30 percent of the total heterosexual copulations are in the missionary position, with the rest taking place front-to-back in the way of chimpanzees, that is, *dorsal-ventrally*. Interestingly enough, bonobos in captivity at the Yerkes Primate Center in Atlanta, Georgia, seemed to prefer ventroventral more than half the time, while bonobos at the San Diego Zoo have a far higher rate for the position. In other words, we can imagine that the position a bonobo prefers for sex will shift according to social context and learned habits.[9] In fact, bonobos often shift positions during the very act itself, as if trying to find the most satisfying position for themselves and their partner. Do these animals really consider the needs of their partners? At the Yerkes Center, researchers videotaped bonobo matings to study such things as pacing and communication during sex, and they concluded that "the speed and intensity of thrusting was visibly altered or terminated as a function of changes in the facial expression or vocalizations of one of the participants."[10]

Sexual contact between females often takes the form of front-to-front embracing and mutual clitoris-rubbing, which I referred to earlier. Kano and his Japanese colleagues describe this style of bonobo sex as *genito-genital rubbing*, an expression usually shortened to *G-G rubbing*. That's an excessively clinical way to speak of an obviously passionate activity. Wrangham and I one day asked two of Kano's field assistants, Norbert Batwafe and Ikenge Lokati, what the Mongandu call this sexual act between bonobo females. Oh, they said, it's called *hoka-hoka*.

Sex between males? Sometimes males assume a quadrupedal stance and, oriented back to back, rub their bottoms and scrota together, an act the Japanese call *rump-rubbing*. Other times they will stroke or rub their penises together, and sometimes—this exciting variation is reported only from Wamba, so it might be a cultural variant among one population of bonobos—they will hang face-to-face from a tree branch while rubbing penises. It's a dramatic behavior dubbed *penis-fencing*. Males will also take the position I described at the start of this chapter, the dorsal-ventral stance common in heterosexual copulation. Kano describes this as *mounting*, and he notes that both males may have erections during the event, although anal intercourse has never been confirmed.[11]

For bonobos, sex produces some clear personal benefits beyond the obvious one of reproduction. Sex is a way to avoid fights and encourage peace. It's a means of reducing tension. You can make friends with sex, and you might very well form useful alliances. Sex is also a desired resource that an individual can trade for other resources, such as food. I'm speaking, of course, of the evolutionary logic of sex. What bonobos *know* about sex is approximately what you and I and most of our fellow humans know about it: that it's a mysterious thing or force associated with strange positions, odd behaviors, and profound pleasures one is inconveniently driven to pursue. Neither bonobos nor humans need to know anything more than that. Evolution plants the desire, and the desire moves us—and bonobos. But you and I, thinking now as evolutionary theorists, can still examine the logic of why evolution has planted the desire. The logic is that sex drives individuals to move their genes into the next generation: That's the direct reproductive purpose of sex. But, as we have seen in the case of bonobos, sex of almost any sort can also increase an individual's reproductive success indirectly, by making peace or reducing tension with others, by cementing friendships and alliances, and by increasing one's potential access to food and other resources.

Many people believe that sex done for purposes other than reproduction is one more decisive mark of the exemptional status of our own species. Bonobos demonstrate the falseness of that belief.

Many people argue that human homosexuality is immoral because it's unnatural. Bonobos show the weakness of that argument. Indeed, bonobos are one of some sixty-three mammalian species known to engage in homosexual behavior, either commonly or sporadically, with the associations either male-to-male or female-to-female, or among both sexes.[12] Among birds, homosexuality has been observed in ninety-four species.[13] So homosexuality is relatively widespread in the animal kingdom, and we can say that for many species it's an aspect of their natural behavioral repertoire. Homosexuality may be common enough among humans to describe it as "natural" for us as well, yet homosexuality is still commonly regarded as immoral and thus is legally banned in almost eighty nations around the world, with the punishments for a violation including death in at least six.[14]

These facts may seem to suggest that the moral rules, often translated into legal ones, are not a reflection of nature but rather of nurture: of society or culture. That's an error of overly simplistic, either/or thinking. The moral rules express both nature and nurture, or, as I more usually describe it, the rules are simultaneously psychological and social.

Sex and sexuality present a realm of social contact so wrapped in significance and fraught with conflict that we can readily predict the development of rules—moral rules—designed to moderate or regularize the nature of sexual behavior for any social species. What we commonly regard as sexual morality, then, consists of a series of psychosocial rules regarding the expression of sex and sexuality: who may mate with whom and under what circumstances. The psychological aspect of those rules, placed by evolution into the minds of all individuals, might be imagined as an embedded map of desire or as a neurochemically-based series of inclinations and inhibitions. The social aspect of those rules appears largely in the form of their enforcements, where socially more powerful individuals press their inclinations and inhibitions upon the socially less powerful. Whether human homosexuality is moral or immoral in any particular society, therefore, amounts to a collective decision made by those with the greater social power.

Is homosexuality moral or immoral? That might be an interesting question, but for me the more pertinent one is this: Why should anyone care? What evolutionary process or event has caused people to have the fears, concerns, interests, and orientations they apparently have about their own and others' sexual behaviors? When I ask that question, however, I discover a tremendous confusion about what those fears, concerns, interests, and orientations actually are. That great confusion is in part a consequence of human diversity.

Diversity could be the single most impressive thing about human sexual behavior and the moral systems that shape it: diversity across cultures and even within cultures over short periods of historical time.[15] The rules and understandings about what is right and what is wrong in human sexuality, we recognize, can differ tremendously from one culture to the next. Take a practice as seemingly simple as covering a woman's body in public. In central Paris, a certain calculated and sometimes bold exposure of the feminine form is one ordinary aspect of everyday fashion—although even here, even in the cultural center of the liberal West, one finds a final limit to the amount of exposure accepted or allowed. In various parts of the Middle East, meanwhile, we discover another cultural extreme, a seemingly more traditional one where a woman in public must be covered fully, head to toe: no exposed hair, no displayed bit of ankle, nothing of the face visible except at the functional opening where someone might notice a pair of shadowed eyes peering out at the world.

Yes, it's important to acknowlege diversity as a significant feature of human sexual morality. Having done that, however, let us now try to peer beneath diversity's multicolored tapestry to identify any features or rules that look steady, predictable, and even conceivably universal among our own kind. I've already mentioned one, which is the practice of sexual concealment. In spite of all the wide diversity in body-covering practices, women are still strategically covered, to some degree, in every part of the world. In spite of what anyone may once have imagined, there's no such thing as a "naked savage," no fully unclothed people to be found anywhere in the anthropological record. Even in those regions where tropical heat and humidity should for utilitarian reasons seriously discourage any body covering, people—men and women—still conceal (or, for some men in some places, carefully fix into position with string) those special parts that we describe euphemistically as the *private parts.* Euphemisms, come to think of it, are a verbal version of sexual concealment. Even in the words we use, we cover. We are inclined to hide our talk about human sexual parts and acts behind a screen of inoffensively vague language—unless, of course, we wish to shock people or to demonstrate our own bold indifference to other people's feelings of modesty and need for decorum. We call the emotions that drive the concealment practice *shame* and *modesty.* True, we sometimes speak of modesty as an emotion or emotional stance that seems to rise and fall with the hemlines; but I'm referring not to a fashionable or superficial modesty but to a fundamental one, the basic motivator that causes men and women of all cultures in all places to strive to keep certain body parts and actions private rather than public, and even to cringe and blush and hide when their efforts fail.

To be sure, it is possible to think of contrary instances where, at first glance, modesty seems readily discarded, while the principle of concealment appears to be blithely dropped. Nudist beaches, for example. But nudist beaches are themselves covered: by a fence, a screen, a screening process to keep out the licentious and the indiscreet. And the trick, the difficult balancing act at nudist beaches, is to display one's body carefully enough to reveal its many wondrous qualities while concealing and inhibiting one's own sexual interests, and, simultaneously, to avoid provoking any unwanted sexual interest among other people. So direct staring is discouraged. Photography may be regulated. Overt expressions of sexual attraction or arousal, such as public kissing or men's erections, will be seriously frowned upon. *Cover yourself* is, I believe, a pro-social inclination and rule, and it is dynamically countered by an antisocial inclination ("Let me see!") that leads to such actions or events as strip shows and pornographic productions. Yet even these uncovered actions or events are ordinarily restricted by an admission price and required proof of majority age, and they are still carried out behind closed doors or otherwise undercover. That's the ongoing battle, the dynamic conflict of self versus others in a moral dyad.

A second universal rule for human sexual morality is embodied in what we think of as the incest taboo. That's the rule prohibiting sex between close relatives: known and expressed everywhere in custom and law. This rule will vary in its intensity of enforcement from one culture to another, and it is nowhere perfectly followed or enforced . . . but then, no rule is anywhere perfectly enforced.[16] Rules exist in a dynamic embrace with the behaviors they forbid; and the existence of the forbidden behavior—in this case, incest—is not evidence for the nonexistence of a psychosocial rule forbidding it.

A third universal rule, one celebrated in popular literature, poetry, and song, and embodied in the customs and laws of human cultures around the world, is the one that promotes pair-bonding. Indeed, the seventh commandment, the only one of the ten that appears to focus strictly on sexual morality, explicitly forbids the people of Israel to violate, with a sexual act, the terms of a culturally reinforced pair-bond arrangement: "You shall not commit adultery."

I use the phrase *culturally reinforced pair-bond arrangement* rather than the word *marriage* not to hammer you with syllables but rather because I'm unsure what an ordinary marriage actually looked like in the time of Moses. People living in contemporary Western cultures sometimes imagine marriage as the romantic union between a pair of heterosexual partners, one handsome man and one lovely woman, who have come together of their own accord, who intemperately promise to cherish each other indefinitely and to treat each other as equals, and who generally agree to follow the seventh commandment requiring sexual fidelity until death parts them. We call that *marriage*, and we often find the idea moving, attractive enough that (when we're reminded of it at weddings) we're surprised by our own spontaneous tears of joy, approval, and identification. Marriage, the version we currently know, is a wonderful and socially important institution.

But even the Ten Commandments suggest a different vision of marriage. For one thing, the tenth and final commandment, the one forbidding covetous thoughts toward one's neighbor's wife, is clearly addressed to men, not women, reminding us that these laws were supposed to be the gift of a male God and were described and interpreted by men, routinely enforced by men, and apparently addressed to men only. Looking back to the Israel of Moses' time, we see a deeply patriarchal society where the power of men over women is institutionalized at every level. Examining the full text of that tenth commandment, moreover, we see that the potentially covetable wife is placed on a list that includes such other covetable items as a house and domestic animals. In short, our contemporary ideal of marriage as a balanced partnership of equals could be a recent development.

The same might be said of our contemporary presumption that marriage is limited to one man and one woman. Although marriage in Moses' time and place seems to have limited a woman to one husband, did it equivalently limit a man to one wife? Probably not, if we take seriously those biblical hints that Moses had two wives and, of course, the references to the great patriarchs indicating that many of them took multiple wives. The famously wise King Solomon seems to hold the record here, with seven hundred wives and three hundred concubines (according to 1 Kings 11:3).

The record of the patriarchs presents an interesting problem for some theologians who insist on reading the Bible literally. Some theological interpreters will say that the Old Testament record provides clear evidence of God's stamp of approval for having multiple wives. Others argue that God actually disapproved of the patriarchs having multiple wives, and that a person can find evidence of that disapproval through further close readings. An evolutionary reading of the same material will be interested not in speculating about God's motivation, but rather in considering that of the patriarchs. What about their own human nature inclined them to take more than one wife once they had acquired a high degree of social and political clout? Some people promote the uplifting perspective that the patriarchs took extra wives because, in an era when so many men died in war, they were generously practicing a form of charity for the benefit of unattached and unprotected women.[17] The patriarchs' supposedly sexual desires, so this theory says, were really charitable ones.

The theory I prefer is simpler. Men take on multiple wives because they enjoy doing so, and they will sometimes follow that enjoyment when they have sufficient social and political power to do so. When, in other words, they get to make the rules. For either sex to take multiple partners is commonly described as *polygamy*—but for humans, at least, it's useful to apply the more specific word *polygyny*, which refers to males regularly forming pair bonds with more than one female.

Perhaps you believe that polygyny is still an exceptional practice, mostly confined to the odd few examples from our historical past. But when we consider the third major monotheistic religion arising from the desert societies of the Middle East, Islam, we find an important and powerful tradition that still allows polygyny to this day among its one billion adherents. Indeed, one survey of the ethnographic record from ninety-three preindustrial societies found that more than 77 percent allowed men to have more than one wife. (Another 6.5 percent gave both sexes the opportunity for plural partners, while 2.2 percent allowed only women to have more than one.)[18] Meanwhile, many happily married men living in Western-style monogamy sometimes risk their reputations and careers, or waste their time and money, by visiting prostitutes or porn sites on the Internet. Since both prostitution and pornography rely almost exclusively on a male customer-base, we might conclude that these two enormously successful industries exist primarily to serve the polygynous inclination. While I will identify the third universal rule in human sexual morality as the one that reinforces pair-bonding, therefore, I should also emphasize that human pair-bonding can take more than one form, including polygyny.

Strategic concealment. The incest taboo. Reinforcing pair bonds. Do these three universals of human sexual morality have any significant parallels among nonhumans?

Strategic concealment must be the oddest of the three, and it might seem most likely to be uniquely human. We place clothes on chimpanzees, after all, when we want a good laugh. Since clothes are so obviously a human cultural product, we might imagine that strategic concealment is strictly a cultural invention among humans: an accident, really, with little or no significance. But there are many other ways of hiding one's own or someone else's sexuality, so clothes are not necessary for concealment; and surely human evolution, not culture, created the modesty urge, which is probably the underlying emotion that helps to keep that covering in place.

One distinctive thing about sexual morality, compared to morality in general, is that the rules are often made and enforced by one sex or the other—and they can affect the general welfare, and ultimately the reproductive success, of the two sexes differently. So when we wonder about the evolutionary logic of concealment, we can begin by asking ourselves which sex makes and enforces the rules. The answer to that question will also tell us which sex benefits the most.

With women's clothing, actually, various human cultures seem to offer two different versions of who makes and benefits from the rules. In the case of women already kept under the strict control of men within a highly patriarchal society, clothes-wearing, the public covering of women, reveals not the power of women but that of men, who use covering as one more way to keep their own women under control while blocking the curious gaze and developing interests of all other men. Alternatively, in the case of women living in a culture where rules about public comportment are much more relaxed, and where they can buy their own clothes with a vision of emphasizing their physical attractiveness, women are much closer to being the rule makers and beneficiaries; and they benefit by increasing their control over the sexual interests of men. They do this not so much by manipulating clothes for concealment as by doing almost the opposite: using clothes to challenge concealment, an act one might call *strategic revealment.* Where women can buy their own clothes and use them freely, to conceal and reveal at will, they have the option of keeping men at a distance—their sexual interests minimized and latent—with one kind of covering. They can increase men's interests with another, more revealing kind of covering. Or, donning some well-crafted and strategically revealing third sort of garment, they can make themselves overtly provocative to men. The ultimate rule of sexual concealment is still in force, to be sure. There are limits to what any human society will allow. But, as these examples show, clothing enables a woman to respect that final limit while at the same time testing or teasing it, and, perhaps, manipulating at will her own level of sexual attractiveness. She does this to get what she wants, and her reasonable goals might range from gaining a useful ally in the office to attracting a desirable mate for the home. That's real power (although you might consider it largely as a response to the power of men), but to use it requires a culture where women can choose their clothes as they will.

Animals are patently unable to avail themselves of such magic; but they can nevertheless create and enforce strategic concealment. Take chimpanzees, for example. You might imagine chimpanzees lack all privacy. You might be thinking that all of chimpanzee life takes place in the open, in full view of all other members of the community. That's the elephant way—but not the chimpanzee way. Chimpanzees ordinarily inhabit the dark and leafy maze of a tropical forest, and they live (like bonobos and humans) in a *fission-fusion* society. While it's true that chimpanzee social life is circumscribed by that single large and territorial community—altogether a few dozen to several dozen individuals inhabiting the same territory—community individuals are, in normal circumstances, virtually never in full visual or social contact with each other simultaneously. Instead, their daily social life amounts to an extended series of comings and goings, groupings and disgroupings, fissions and fusions. In that sense they live very much as we humans do: going about our daily activities from one task to the next, gathering together with several friends at one moment, moving off alone to shop at the grocery store at another, meeting up with a family member or accidentally running into a best buddy at yet another moment.

Of course, certain events can draw a noisy crowd. Some chimpanzee catches a monkey, and everyone else in the vicinity moves in to beg, squabble, or connive for a piece. Another exciting event is when a female approaches ovulation: that all-important moment of fertility in her thirty-six-day cycle. No chimp knows a thing about fertility, but they all notice when a female's body announces fertility's approach and arrival. It's the time when the skin around her external genitalia turns bright pink and becomes markedly swollen. The males seem to find such a development irresistibly pornographic. The female becomes at once sexually active and attractive in the extreme, and she will soon be followed by a noisy scrum of aroused males. This is social excitement verging on chaos, with the males posturing, displaying, and sometimes openly fighting over the matter of who and when. Chimpanzee sex has often been described as *promiscuous*, and, indeed, ultimately several males will mate with the one fertile female. Still, this seemingly chaotic situation favors the alpha male and his political supporters, since they have the social power to insist on mating when the female is maximally swollen and most sexually attractive—and also most likely to conceive. In other words, their social power translates directly into reproductive power.[19]

However, less dominant males can sometimes arrange a second scenario for their own benefit. Sometimes a male will find himself near a female who is at the start of her fertile period but has not yet attracted the usual crowd. It's just him and her, two ordinary chimps foraging alone together in the woods, and he arranges—through gestures, threats, and, if necessary, a pummeling or two—for them to make themselves even more alone by slipping away from the usual haunts and finding a more remote part of the forest. In this fashion, they enter into a *consortship*. For a few or many days, he and she have entered a situation where the sex is quieter and less frantic, and where this one male has exclusive access to this one female.

Chimpanzee consortships might superficially appear to offer a better deal for the female than the more standard, promiscuous variety of sex. The potential for her and any of her dependent offspring to be hurt by the occasional fights that can break out among the males is less. But from an evolutionary perspective, a female's ideal sexual partner would probably be the alpha male, the one who predictably wins that promiscuous competition—and probably not the crafty but lower-ranking male who has drawn her away from the competition altogether. The alpha, after all, is most likely to give her a son who grows up to become alpha in the next generation; and that evolutionary goal will, for her, have been translated into an emotional goal. Think high school: She spontaneously finds she would rather go to the prom with that swaggering, self-confident football team captain than with the manipulative and secretive second-string member of the drama club. In any event, like much of what happens in the patriarchal world of chimpanzees, this consortship has predominately been arranged by a male for his own benefit. He has created and enforced a temporary state of quiet monogamy by strategically concealing her, and her fertile condition, from the other males.[20]

The incest taboo. If our clever male finds his advantage in concealment, however, this fertile female may seek her own advantage in revealment. The social power temporarily accruing to females when their private parts become decidedly unprivate—turning bright pink and swelling up like a balloon—are well-established. She becomes more popular. Males become temporarily more attentive and even deferential. But perhaps the most critical aspect of that power becomes evident during early adolescence, at the critical moment when a female is likely to emigrate from her birth community and try integrating herself into the society of the neighboring community.

This amounts to running away from home, and it's only possible because the female's vivid flag of sexual receptivity also serves, for the neighboring males, as a compelling flag of peace. For her to leave her birth community, after all, requires crossing the territorial boundary and thus subjecting herself to the potential of a crippling or lethal attack from those patrolling and xenophobic neighboring males. No male would ever think of making this same difficult journey by himself. Nor would an older female. But this early-adolescent female seems to recognize that her state of attractive availability has temporarily changed the rules. And indeed, as she soon discovers, the usually violent responses of the neighboring males soon turn in much kinder, gentler, and more intimate directions. Yes, she is welcome.

It's clear why she *can* do it. Her revealed sexual state allows the gates to be thrown open, the barriers let down. But why should she *want* to? What is the underlying reason why evolution has planted this desire to emigrate in her psyche—planted it deeply and consistently, and done so not only with her individually but with her full sisterhood of chimpanzee females? Many of the females, upon reaching early adolescence and their first real flowering of sexuality, will emigrate from their birth community. Males never, never do. This same pattern, incidentally, is also followed by bonobos. Males stay in their birth community, while females routinely leave. And it is found in reverse among other species, such as the olive baboons living at Gombe Stream National Park in Tanzania, where Jane Goodall's chimpanzees also live. Among the olive baboons, females stay in their troop of birth, and the males emigrate.

Probably the underlying evolutionary reason for this larger pattern is to avoid the likelihood of individuals mating with their close relatives: incest avoidance. Obviously, none of these animals know the name of what they're doing or understand the logic of it. They don't have a language to discuss these things, and they don't have a direct way of examining the concept of *close relative*. But they don't need to make such fine distinctions. They need to know only whom they have grown up with or spent a good deal of close and personal time with. Knowing that is enough, in most natural circumstances, for individuals to distinguish, indirectly, a close relative from others. And it is enough for evolution to design a system that will avoid the serious genetic harm that can result from the mating of close relatives.

Yet, these emigrations seem so fundamental and automatic, so deeply embedded, so . . . instinctive. Where are the deliberate behaviors and choice-making that we expect to find in moral behavior? Where are the rules? Who does the enforcement? I might say that the emigrating chimpanzee female is making rules against incest and choosing to enforce them by fleeing her birth community. But I would also like to point out a second variety of incest avoidance among animals, where the rule-making and enforcement are more active and apparently deliberate.

For the chimpanzees at Gombe, matings between adult sons and their mothers are "extremely uncommon," while sex between brothers and sisters has been observed only "very infrequently."[21] Young males do develop a sexual interest in older females—and why not? But if the older female is one's own mother, things are very different indeed. At Gombe, the brothers Figan and Faben were regularly seen in the company of their mother, Flo, when they were sexually mature and she sexually attractive and active.[22] During those times, every male in the community either mated with Flo or tried to—except for the two sons. Meanwhile, Faben and Figan's sister, Fifi, mated with every single adult and late-adolescent male in her community during her first fertile period—everyone, that is, except for her two brothers. Figan, who was thirteen at the time, didn't seem interested in Fifi's sexual state, while the nineteen-year-old Faben twice approached her with his hair raised in excitement and his penis erect; and she, screaming at the very sight, ran off.[23] In later years, both brothers tried to mate with her at times, but she regularly resisted their advances. When they did succeed, it was often clear that one or the other had coerced her in spite of her active resistance, adding to the more general impression that females, not males, actively enforce—or try to enforce—this particular rule.

Reinforcing pair bonds. One interesting thing about bonobos is that their female emigrations, although superficially resembling the case of chimpanzees, challenge male power while making use of female power. Female power is the biggest of the bonobo surprises, and it comes in large part from female solidarity. That solidarity relies, in turn, on the females' capacity and inclination to form strong pair bonds.

An adolescent bonobo female emigrating into a new community will probably seek out a powerful resident adult. Another female. Moving at first with circumspection, shy glances, tentative approaches, an attempt at grooming, she wins the interest and then, seemingly, the affection of this older female, and she acquires thus a mentor and special friend. The older friend becomes a sexual partner—a lover, we might say—and, also, since she is supported by her own strong affiliations with the other females, a significant protector. Through this meaningful relationship, solidified by emotional and sexual bonds, the new female is accepted into the community, and with the force of female solidarity behind her, she becomes an important individual who may well become dominant over some or several or all of the males. Females form the most significant pair-bond relationships among bonobos, and the network of female relationships helps stabilize the larger social system.[24]

Clearly, you and I can recognize that it's a far, far cry from the active and often promiscuous couplings of these wild and hairy apes to the structured and culturally sustained matings and relationships of humans. Marriage is a program for our own welfare, so many people say, that has been designed by God himself to endure for a lifetime. Most people, it appears, earnestly hope to find a special relationship that does indeed last a lifetime. Such special relationships are endorsed by our own emotions, reinforced by our cultural institutions, and additionally protected by cultural systems from the undermining effects of infidelity: thus, the seventh commandment prohibiting adultery. Our most important pair-bonded arrangement, we tell ourselves, should be exclusive.

Exclusivity of that sort, it is true, will not be found among bonobos or chimps. For exclusivity and for a stable, even a lifetime's, close association, emotional connection, and shared responsibility, we need to look elsewhere. Gorillas, for example. Gorillas live in family groups that consist of one fully grown adult male, one or a few or several adult females, and their young and growing and nearly grown offspring of both sexes. One father, one or more mothers. That's polygyny; and people often say, as a reasonable analogy to life among us humans, that gorillas live in harem societies.[25]

As Herman Melville reminds us in *Moby-Dick*, a powerful kind of competition sometimes takes place between males over the question of who gets to mate with a particular female. It's the male-against-male competition that we can find among male sperm whales fencing with their long lower jaws. Or in the fighting of two bull elks who warringly interweave their antlers. Or in the dueling of nineteenth-century gentlemen driven into lethal competition by one desirable woman. This sort of male-against-male competition occurs among so many mammal species so often that people have come to think of it as a cliché. But biologists see its deep importance among certain species because the predictable victory of larger, stronger males (along with a female preference for such victorious males) results in the evolution of a striking sexual dimorphism: males bigger and more powerfully built than females. History is recorded in the body. Male elephants, for instance, are twice as big as females, so we can conclude that male-against-male competition for mating rights must have been very important in this species' evolutionary history—as it must have been among gorillas and orangutans, where males also are twice the size of females. Among chimpanzees, bonobos, and humans, the sexual dimorphism is less extreme.

Melville declares, with a confident and possibly ironic flourish, that they do it "all for love." But do they? What is the emotion that drives such serious and potentially damaging competition among males? We often presume that only humans will fight and die for love, true love, while what encourages animals to mate and sometimes to pair off is a much cruder and simpler instinct called *lust.* We say we're being our best human selves when we love, but we're expressing our unfortunate animal selves when we give in to lust: that is, have sex without personal attachment. That's one common idea, but I prefer the concept advanced by anthropologist Helen Fisher in *Why We Love* (2004), that people and many animal species pair off following three separate though often associated emotions, or (to use her language) "three primordial brain networks that evolved to direct mating and reproduction."[26] The first she calls *lust*, which is the urge to mate generally. The second she identifies as *romantic love*, which is the elevated sense of special attraction to a particular individual. The third is *attachment*, which is a more lasting bond between mates that encourages them to stay together and jointly care for their offspring.

Obviously, lust is important for evolution since sex results in reproduction. But particularly for those many species where the young are born incapable of surviving independently and thus require parental care—nursing, protection, instruction, and so on—the second and third emotional complexes might be significant indeed. Among mammals, mothers produce food and nurse their young; and for many mammal species, mothers do all the parenting. This is true for our closest relatives, chimpanzees and bonobos. The males don't join in parenting and often don't even seem to know who their offspring are. But when we look to those species with pair bonds strong enough for both parents to join in raising, feeding, and caring for their offspring over several weeks or even years, we have to wonder what motivates both parents to stick around and share the parenting. What emotions draw a pair of animals to identify each other individually, and to value each other as individuals in a sustained enough manner to account for that long-term shared investment in territory and tasks? They are the emotions or emotional complexes identified by Fisher as romantic love and attachment.

A couple of years ago, I spent part of an afternoon watching—through binoculars and from a distance of about forty yards—two lions mating. They would begin with both lying on the ground, he partially behind her and to one side. Then, as she continued lying on her side, the dark-maned male would stand up, squat over her, mount, enter, and thrust. He seemed to purr, deeply, as he did these things, while at the same time she cough-roared rhythmically. After a few seconds, he lay back down, and she turned over onto her back, exposing a white belly and rolling with all four feet up for twenty or thirty seconds. Then she turned back onto her side, her tail twitching once, twice, thrice. Two minutes later and they were at it again. This time, he roared briefly as he thrust; and she, again, cough-roared. A couple of minutes passed, and they did it a third time, accompanied by the same noises, and again she rolled onto her back and twitched her tail. Moses ole Sipanta, the Maasai guide who had brought me out there, said that lions were known to have sex like that up to a hundred times a day. That number may not be an exaggeration, since the official record for lions is 157 times in fifty-five hours.[27] But in any case, I was impressed by such fierce endurance and surprised by such steady ardor. Instead of the dysphonic, back-alley caterwauling of domestic cats in heat, these big wild cats relaxed, took their time, made possibly affectionate and certainly pleasurable noises, stuck around after coition, and turned back to do it again.

It seemed pleasurable. It was steady. It looked affectionate. Was this evidence of romantic love between two animals? Not necessarily. According to Fisher's definition, we need to know more. We need to discover if these two lions had chosen each other as particular individuals, and we would also want to observe any courtship behaviors that served to encourage that choice and specificity. "Animals love," Fisher insists, and the emotion shows itself as being "drawn to specific others." To be sure, animals don't analyze and write about their feelings, as we do, but they do express them with the states and behaviors we see among people who claim to have fallen in love. "Temporarily charmed, these lovers step to a universal beat, croaking, barking, flapping, trilling, strutting, staring, nuzzling, patting, copulating—and adoring—their preferred mating partners."[28] They become energized and excited by that special individual. They stop eating. They lose sleep. They show signs of euphoria and obsession, and they behave affectionately and sometimes possessively. The feelings and behaviors associated with animal romantic love, Fisher adds, are moderated, as in people, by the same two ordinary chemical compounds of the brain, dopamine and norepinephrine, and they involve comparable parts of the brain.[29]

So romantic love is particular. It presumes choice among individuals, and that choosing will be stimulated by courtship and consummated in mating. But Fisher's third emotional complex, attachment, working through a third neurology and biochemstry, establishes a bond that lasts beyong the mating phase. In this third case, the mated partners maintain their relationship, build a home or den or nest together, and raise their young: the sort of monogamous pair-bonding we find among comparatively few mammals. Birds, on the other hand, are well-known as monogamous creatures, with nine out of ten bird species forming paired partnerships. Birds, so people say, mate for life—or at least for the duration of a breeding season.[30]

Birds can become emotionally attached to specific partners, set up households, and begin having offspring; and they may care for those offspring together as nurturing parents. That's comparable to Fisher's ideas of behavior driven by romantic love and attachment. Birds, however, often find that such romantic love and attachment are not enough to ensure a perfect union. Sometimes a bird will be distracted by another motivating system—we might call it lust—that can lead him or her to be momentarily drawn to another bird who is not the regular partner. Even birds live in a social world where some values will come into dynamic conflict with others.

Take the case of black-capped chickadees living in eastern parts of Ontario, Canada. These lovely little creatures go through an autumn and winter social period when they congregate as stable flocks consisting of around a dozen birds of both sexes. During the social season, the flocked birds form male and female dominance hierarchies: relationships expressing a sense of who is more powerful and important than whom. The males form one hierachy among themselves. The females square off among themselves. When spring arrives and the social season ends, the males and females then pair off with each other, still keeping in mind their individual degrees of social importance. The more dominant males form pairs with the more dominant females. The less dominant males form pairs with the less dominant females. And so on. In this fashion, each pair begins a monogamous relationship. They court. They mate. They jointly work on their nest and move into it. Together they feed and raise little chickadees while defending the nest and its surrounding territory from incursions by others of their kind.

All is well, then, in the suspended suburbia of the black-capped chickadees . . . except that a female will sometimes sneak off to mate with someone other than her partner. And when the breeding season ends, the offspring have left the nest, and the Earth continues to turn around the sun, so comes another social season. The couple returns to communal life in the flock, and the females and males once again have the opportunity to consider others at length and up close. If a female finds a male who appears to be a better catch than her old male of the previous year, she may just divorce the old guy and start a new household with this new and better, probably more dominant, male.

Dominance marks who gets a contested resource, so we can imagine that identifying the relative dominance of males is important for the female, who looks to find the one who will be the better protector and provider. But how do we know that the female black-caps actually prefer the more dominant males? An experimenter once temporarily removed seven high-ranking females from their seven high-ranking male partners, then watched to see what happened to the now available males at the start of breeding season. Before long a number of nearby females, living in monogamous relationships with lower-ranking males, simply left their old partners and moved in with the better males. This was wholesale mate abandonment. This was flighty females choosing higher-ranking males. But the situation was reversed once the original seven females were brought back to the area. The original females found their old mates and chased away the opportunistic females, who then returned to life with their original, lower-ranking partners.

To test these things from another angle, the experimenter next removed six low-ranking females who had partnered up with six low-ranking males. What happened? Nothing. None of the remaining females in the area left their partner to nest with the newly available but low-ranking males. Why should they? These males were the losers of the flock. Thus when the six females were finally returned to the area by the experimenter, they also returned to their original low-ranking mates—easily done because they didn't have to chase away any bold interlopers.[31]

When we look at monogamy among black-capped chickadees, we're seeing an imperfect union. There are those occasional sneaky infidelities. There is the possibility of mate abandonment should a better mate become available. And there is always a significant chance of divorce at the end of the year. However we describe the internal system that makes these charming creatures do what they do, we know that there will be unpredictable events and sometimes choices to make. Their monogamous system is not perfectly stable because these birds also have a nonmonogamous tendency. To some degree, that tendency is countered by individuals who enforce monogamy, albeit in the simplest possible way: by defending territory, by chasing away any opportunistic interlopers.

Only around 15 percent of primate species are monogamous;[32] and among that straitlaced minority are those known as the lesser apes—or gibbons. Taxonomists list about a half dozen or more gibbon species, all of them living in the fast-disappearing tropical forests of southeast Asia. The species I'm most familiar with can be found on Indonesia's Mentawai Islands. These are the last discovered and least known of all the gibbons, but also the ones I visited twentysome years ago. The local people call them *bilou*, while scientists prefer to talk about *Symphalangus klossii*; and it was reading the scientific reports that finally inspired me to go see these marvelous creatures for myself. From the big island of Sumatra, I arranged an informal passage on a freighter traveling out to the much smaller island of Siberut, the largest of the Mentawai Islands. Then I took a motorized dugout canoe upriver, into the island's rain-forested interior, stopping finally at the village of Rok-Dok. There I was welcomed by a handsome and generous young man named Karlo Saddeau and was given a place to sleep inside his stilt-elevated, split-bamboo home, alongside his wife, several children, and aging parents. From the front porch of Saddeau's house, cup of tea in hand, I could listen to the lovely haunting songs of bilous as dawn broke through the night's mist; and finally, one day, I was able to find them out in the forest. I saw small, dark-haired primates with long arms who were swinging—or perhaps flying—from branch to branch and tree to tree with a dazzling grace and speed.

Gibbons are famous for their acrobatic prowess, but they're also well-known for their monogamous lifestyles. Males and females pair off, locate a home tree, defend a home territory—for bilous this would be about sixteen to seventeen acres—and jointly raise babies until they are old enough to leave home. As the pair-bonded gibbons are doing that, they maintain their monogamous ways. How? What makes them behave monogamously when, clearly, there are temptations for both males and females to violate that monogamy? Gibbons sing in the mornings, which announces the position and, roughly, the dimensions, of a monogamous couple's territory. And then, at those distressing moments when another sexually mature gibbon appears and crosses or threatens to cross into their territory, they chase him or her away. More specifically, males chase away intruding males, while females chase away intruding females.[33]

The gibbons' rule? *Don't mess with this happy family.* The enforcement? The male and female gibbons each try to prevent any distracting sexual competitor, tempter or temptress, from making it very far into the family territory.

Romantic love is such an appealing and essential emotion for humans—and really, where would we be without it? I will eagerly join in the chorus, placing my own voice alongside that of Frank Sinatra, B. B. King, Mick Jagger, Elvis, Cher, Aretha Franklin, and a hundred other top entertainers. We all celebrate this emotion; and we should all celebrate the longer-term attachment that sometimes follows romantic love. After all, these two emotions are not merely important for our own individual happiness. They are fundamental for ensuring the stability of our households and, often, the protection and comfort of our children. But romantic love and attachment are limited by their inherent instability, and they are also sometimes countered and complicated by that third emotion, the one Helen Fisher calls lust.

To learn more about lust and its contribution to the human drama, drive on down to your nearest country-and-western watering hole. Put your quarter in the jukebox to hear, modulated by steel guitars and a smoky beat, an adenoidal lament on a theme of gonadal misadventure, a sad story of what really goes on in this sad, sad world: cheating mates, broken hearts, and drinks on the rocks to ease the pain of a marriage on the rocks. This is what happens when love and attachment come into conflict with lust. A person can feel helpless under the sway of these three powerful and sometimes conflicting emotions, and it is possible to drink that drink, have another, and do nothing more than sit at the bar and slowly review the literature of deceit, betrayal, and loss. To nurse one's broken heart. To anesthetize the painful memory of broken love and severed attachment.

Blot the memory: That's one option. Another is to listen to the voice of a fourth emotion and, perhaps, act in response. This fourth emotion is called *jealousy*, and the story of its pains and powers will be radiating from the same jukebox of the same establishment. It can indeed motivate a person to take serious actions that might actually fix the problem. Or make it worse. This fourth emotion can suddenly bring you back onto your feet and striding right out of the bar, slipping back into your automobile, and intemperately driving off to locate the ugly opportunist who has stolen your love and broken up your marriage. Perhaps, you think, you can find that person, drive him or her away, and win back your true mate. It might work. Maybe your beloved will even be moved by the sight of you in all your masculine (or feminine) arousal, vigorously and vociferously declaring once again the intensity of your love while ridding her (or him) of this dastardly interloper.

We have no single word to describe the behavior you're considering, at least when people do it, although it commonly and often disastrously happens. Sometimes it even seems appropriate and may actually, when viewed in a certain light, appear romantic. That's the positive script followed by young Dustin Hoffman playing Ben Braddock in *The Graduate*, when Ben breaks into the church a critical few seconds before the I do's are done, boldly racing up the aisle while bellowing out the name of his one true love, Elaine, sincerely grasping her hand to wrench her away from the insincere clutch of her false love. As I say, we don't have a particular word to describe this behavior among people. When animals do it, however, biologists call it *mate-guarding*. One has a mate. One vigorously and vociferously guards that mate from skulking interlopers.

Few people have thought to name the emotion underlying mate-guarding among animals, although with humans in similar situations we recognize it as *jealousy*. Mate-guarding is the behavior of that female black-capped chickadee who, returning to her nest, drives away the cheeky female who moved in during her absence. It's the act of a male Mentawai Islands gibbon threatening the other male who has ventured too close to the home tree, or of the female gibbon who drives off her own sexual competitor. Mate-guarding enforces already established pair-bonds; and the enforcement might be said to support a rule like *Don't disrupt this happy nest.* Or *Don't cross the threshold of this territory.* Or *Do not threaten this established bond.* Of course, only among language-using humans is the rule actually written down and talked about—and, having been written down and talked about, more broadly supported by social sanction and threats of supernatural displeasure and punishment: "You shall not commit adultery."

The seventh commandment should help to save us from ourselves, our own conflicted selves; but it might also save us from the undermining behaviors of others. If enforced well enough through social and psychological means, the rule can help to preserve our romantic loves, our long-term attachments, the integrity of our families and communities—and perhaps most significant, it might help us avoid that potentially dangerous situation where we feel compelled to enforce the rule ourselves. Personal enforcement means individual mate-guarding, which means that the pathetic, drunken lout in the bar has actually climbed into his car, revved the engine, and is now on the way to creating chaos while believing he can re-create order.

CHAPTER 8
Possession

Thus the most vexatious and violent disputes would often arise between fishermen, were there not some written or unwritten, universal, undisputed law applicable to all cases. . . .

I. A Fast-Fish belongs to the party fast to it.

II. A Loose-Fish is fair game for anybody who can soonest catch it.

—Herman Melville, *Moby-Dick*[1]

A dolphin catches a very nice fish, but instead of eating that sweet morsel right away, he or she starts playing with it. The dolphin throws the fish into the air, six to seven feet above the surface of the water, then swims over to grab the fish and toss it once again, continuing to toss around and play with the stunned creature. Other dolphins are there. They may be very interested in the fish. They may go right up to it, investigating closely. But they don't touch it. Sometimes the fish is a real prize, and the other dolphins get sexually excited and start mounting the dolphin with the fish. But they don't try to take or eat the fish; and the dolphin who has tossed, played with, and otherwise displayed that prize at last casually swims over to where it's floating on the surface, just swimmingly saunters over, and eats it.

Marine biologist Richard Connor, who has studied bottlenose dolphins for the last twenty-five years at Shark Bay in Western Australia, made a video of the most extreme case of this behavior he's seen. A dolphin known as Hai caught a Spanish mackerel, a rare species for that area, and the fish obviously was a prize. The other dolphins were excited, and all Hai did for ten minutes was display his catch. The dolphin lay on his side, holding the fish with his mouth around the head, thus showing off a long stretch of the unusual fish's body. The other dolphins were racing around Hai, porpoising over him, and mounting him. But none of the other dolphins touched his fish. They could easily have taken it. It was right there, and they were all around. They didn't touch it. So it seemed as if everyone had recognized that Hai owned the fish.

Hai is a big, tough male, but this may not have been a case of everyone's being intimidated by Hai's ability to hold on to that fish through sheer force. Smaller dolphins will do the same thing, Connor tells me, and female dolphins will do it when the bigger and more dominant males are around. Juveniles, too. Nobody messes with that fish because, so it would seem, the dolphin who possesses the prize has succeeded in claiming ownership of it, and the other dolphins recognize it as owned property.

I don't imagine that bottlenose dolphins think deeply about ownership or intellectually understand the concept of private property. I simply imagine that when a dolphin or some other animal in a social group possesses a valuable item, such as a good piece of food, a predictable situation of potential serious conflict has been created. The other members of the group will desire that same object. And for many species, I believe, evolution has produced a system for mitigating that potential conflict by establishing psychosocial rules making ownership and property possible.

Possession is a physical act. Ownership is a social condition. Property is the consequence of that social condition.

We humans deal with property and the rules of ownership every day of our lives. Property, we know, can be an object, a piece of real estate, or sometimes in some historical circumstances another person. People own property, but property can also be rightly or wrongly possessed. Banks hold cash as owned property. Bank robbers steal that cash, but although they can possess it, they will have trouble convincing anyone that they legitimately own it. So property includes, as an inherent aspect of its definition, an understanding of certain psychosocial rules regarding ownership, and we know what those rules are—the primary one being identical to the eighth commandment as written in Exodus 20:15: "You shall not steal."

The rules of ownership apply to every sort of human community and to all kinds of human endeavors—and for nineteenth-century whalers working on the high seas, so Herman Melville reminds us, the most important piece of property they could hope for was a whale. The value of a whale as property might be marked in dollars and cents by a ship's accountants onshore, but for the working whalers at sea that value could also be imagined in sweat and blood and, often, as a loser's trade for the bright gift of life itself. A whale was a real prize.

Also, a slippery one.

Imagine, for example, you've spent a full day on heavy seas in that small boat chasing one big whale. You and your shipmates darted him with a half dozen irons, yet he's still alive, has still avoided that final thrust of steel to the heart. He must be mortally wounded, but as darkness approaches and you return to the ship, this animal is still swimming. Next day, you see another ship with a giant prize fastened alongside. It's a dead whale, but one, as you see upon closer inspection, with your own specially marked harpoon sticking out of the side. This was the very prize you risked your life for. Your effort and your harpoons, and those of your shipmates, killed that whale, and the other ship merely gathered in a shark-pocked carcass. Outrageous! What are your rights of ownership in the matter?

In the *Moby-Dick* chapter called "Fast-Fish and Loose-Fish," Melville's narrator describes the rules concerning whales as owned property. For the purposes of human commerce, a dead whale is a big fish existing in one of two possible states: fast or loose. A fast-fish may have been made physically fast to an occupied ship or boat with any conceivable means of connection: be it a rope, a chain, a line, a violin string, or a filament of horsehair. Or a whale can be made symbolically fast if someone plants a pole with an identifying pennant, called a *waif*, into the floating carcass. A fast-fish, legitimately claimed by line or waif, is now owned by whoever made the claim. Claiming the whale formally, as a social event, then, gives someone ownership, and that ownership inherently includes moral and legal support. Morally, anyone trying to take that whale is doing wrong, is violating the eighth commandment against theft, and can expect enforcement to come from the combined forces of social and supernatural disapproval. Legally, one's possession of the whale will be sustained by the rough machinery of organized justice. Moral and legal protection arrive the minute a whale becomes a fast-fish.

A loose-fish is a whale not claimed by line or waif and therefore still open for the taking. Alive or dead, the whale is no one's owned property and fair game for anyone who can make a claim.

Many animals regard particular objects as special, but special objects are not the same as owned ones.

My dog Spike is strangely attracted to tennis balls, and upon discovering one in the park, she will pick it up in her mouth and not let go. It must be uncomfortable, carrying such a dry and tightly napped object in your mouth, but Spike does so with a long-faced solemnity, as if the universe demands that she pick up tennis balls. She'll bring it home and lie, sphinx-like, on the kitchen floor, her latest acquisition positioned for further consideration between two front paws.

Smoke has never really cared about tennis balls. Tennis balls are not her thing. She would much rather carry around an old turd.

I pry open Smoke's clenched jaws to remove the disgusting object, and I will soon dispose of Spike's slimy orb as well. Both resist, but neither is capable of resisting for long, and so their special objects are soon forgotten. After all, where would they keep them anyway? Nowhere I can't get to. Most animals have a personal storage problem, the same one people must have contended with before the invention of clothes with pockets, houses with doors, and doors with locks; and the lives of wild animals are ordinarily marked by a severe economic simplicity. They spend their days finding and eating food, using or needing few special objects.

Anthropologist Richard Wrangham, who studies chimpanzees living in Kibale National Park of western Uganda, recalls an instance of one young chimp becoming attached to a special object. At eight years old, Kakama was still young enough to spend almost all his time alongside his mother, Kabarole. Kakama was an only child, but his mother was pregnant with another. Perhaps it was boredom, the stultifying combination of a sluggish mom and no playful siblings, that led young Kakama one day to begin playing with that small piece of log. Wrangham watched as the young chimpanzee dropped facedown onto the log, wrapped his arms around it, and began rolling around with the object clasped tightly in his arms. When his mother rose up and began to leave the area, Kakama picked up the log and dragged it along behind him.

Kabarole climbed high into a tree to reach some ripe fruits. Her son followed—laboriously, however, because he was dragging the log. When they were finished feeding in that tree, they moved on to another, climbing into three more trees altogether that morning, with little Kakama still hauling along his new toy. He would carry it on his back, drag it on the ground, balance it at the back of his neck, apply it like a crutch or an extra leg. One time, the object fell from a branch thirty feet down to the ground, but when he and his mother had finished feeding in that tree, Kakama scurried over to gather up his newfound toy before racing back to follow his mother.

As darkness approached that evening, Kakama climbed into a tree and wove a few branches into the usual sort of sleeping nest, not far from the nest his mother had made for herself, but Kakama had brought his log with him, and as the researcher could see, the young chimp was now lying on his back inside the nest, all four limbs in the air, and holding the log above him with both his hands and feet. He seemed to be playing "airplane" with the object, much in the way that chimpanzee mothers do with their babies. Then Kakama made a much smaller nest, a toy nest, and after the chimp had finished, he placed the log inside the nest, like a doll in a crib—then climbed in himself.[2]

The example of Kakama suggests some of the intelligence of chimps. It's relatively rare for wild animals to use special objects as toys. Less rare is using special objects as tools. Elephants may apply a stick to an itch behind an ear. Certain kinds of birds use sticks as probes. Chimpanzees in parts of East Africa will take a stick or a piece of palm frond, trim it down until it's the right shape and length, then, carrying the modified object over to a termite mound, use it to fish for termites. Chimpanzees also use stone tools, a West African tradition found among those chimps living west from central Ivory Coast into Guinea and Liberia, as far as the Moa River in Sierra Leone. The apes wield stone hammers (sometimes wood ones) onto stone anvils (sometimes wood) to harvest the edible innards of six different varieties of hard-shelled nut.

Chimps living in the Taï Forest of Ivory Coast traditionally crack open and eat the nuts of African walnut, gray plum, and *Panda oleosa*—and the *Panda* nuts are notoriously hard to crack. To break open *Panda*, the Taï chimpanzees use only stone hammers, big ones weighing up to forty-five pounds. If you should come across a chimpanzee hammer in a West African forest, you might be excused from imagining that you've just found a human artifact. A good stone of the right size and shape is hard to find in these forests, so when a chimp does find one, it gets used over and over, making the hammers artifically rounded by months or years of repeated use. So the hammers are special objects. The anvils are special, too, since it's also hard to find the right stone for this task. The anvil should be of a certain size and hardness, and it has to be flat. They also are artificially modified by long-term use, acquiring deep grooves and indentations.

The nut-cracking is an extraordinary thing to see. The chimps, moving in small or large groups through the forest, come upon one of their favorite nut trees that is dropping ripe nuts, and suddenly you hear the sound of cracking, and you see—over here—a young ape gathering an armful of nuts, walking upright on his hind legs to an anvil, sitting down there, picking up the hammer that lies nearby, and industriously cracking open nut after nut. Over there, you watch a mother squatting next to her young daughter. The mother cracks open a nut, uses a twig to dig out the meat. The daughter watches intently, and after a while she places a nut on the anvil and gives a try herself: mightily lifting the hammer and dropping it on the nut. There could be a half dozen hammers and anvils at this scene, all being used, and so you're listening to the sounds of an industrial sort of nut-cracking: *tap-tap-tap-tap-thwack, tap-tap-tap-tap-thwack, tap-tap-tap-tap-thwack*. Once these chimps have finished cracking nuts and are ready to move on to the next promising food source, they leave the hammers and anvils behind, conveniently placed in the forest right beneath the nut trees, ready to be used again once some more nuts have fallen to the ground and another group from the same community shows up to harvest them.

The hammers and anvils are special objects but not owned property: They're forever loose-fish, never fast-fish. . . .

Here we might imagine the enticing vision of chimpanzees living in a Marxist paradise, a world where the demon of private property has been exorcised and replaced by the angel of communal sharing. Comradeship among the apes. Rampant egalitarianism among the apes. Or maybe the demon of private property hasn't even appeared in the first place. Are chimpanzees simply less greedy and nicer, or more egalitarian, than most people living in modern societies? Have these animals not reached an advanced enough social or cultural state, or a high enough level of intelligence or facility with language, for ownership and private property to exist? The old capitalistic principles of supply and demand make these objects special. The supply is low. Only a few good, right-size and properly shaped nut-cracker and anvil stones are around. The demand is, periodically, high. When the chimps arrive at a place where the nuts have recently fallen, lots of them try to eat this important food, and they need those hammers and anvils to do so. Low supply interacting with high demand combine to make them special objects, but they are still not possessions with the potential for becoming owned property.[3]

Why not? Stone hammers and stone anvils are simply not, for chimps, pragmatically possessable. After all, what chimp in his or her right mind would even think to drag around a forty-five-pound rock? If the supply were low enough and the demand high enough, yes, I can imagine a chimp trying to hold on to a hammer or an anvil. But not for long. An ape cannot live on nuts alone. Sooner or later, he or she has to move on and look for other foods, and the hammers and anvils are just too heavy and too awkward to carry along.

For wild animals, food is often low in supply, high in demand, and possessable—and so eating, just gobbling it down, might be described as a significant act of possession. But taking possession is still not the same as claiming ownership: of defining and protecting an object as yours. Taking possession is a private and often solitary act, whereas claiming ownership is a social act that becomes important when one's possession is, for some reason, delayed or otherwise imperfect or incomplete and therefore likely to be contested by others. Ownership, then, is successful possession in a social situation.

For chimpanzees, only one kind of possessed object becomes owned property, and that is a piece of meat: the carcass of a prey animal just killed and about to be eaten. Meat in these circumstances is low in supply, high in demand, and can be possessed. But because raw meat is also tough and chewy and takes a long time to eat, its possession can and will often be contested.

Sometimes a lone chimpanzee will catch a monkey, a small antelope, a baby wild pig perhaps, then slink off into the bushes and quietly eat that delicious meal all alone. That's a private act of possession. But most of the time, meat-eating is a public act of some duration, with a good deal of competition over possession and therefore a need to claim, in any way possible, ownership. Chimpanzees often hunt cooperatively, with two or three or four adults actively pursuing a single prey. It's impressive. The chimps may stalk and silently position themselves at different spots in the forest around their unsuspecting prey, and then, as if on signal, it begins. One chimpanzee will chase and drive the surprised and now fearful animal in one general direction, which is right into the hands and teeth of the other two or three chimps, who are positioned at the most logical escape routes. What started in stealth and silence is now transformed into a violent cacophony of wild screams, barks, hoots, until the captured animal, crying out in panic and desperation, dies in a whirling world of violence, pain, and noise. Meanwhile, every single chimp in the vicinity, having heard the sounds of a hunt and kill, converges. There's meat!

That just-killed monkey, that meat, will first be possessed by the individual chimp who captured and killed the animal. He or she caught and killed and thus made fast a loose monkey, and he or she is now mightily trying to keep it fast: to own it. There may be some quick sharing of the meat, and perhaps we can imagine that a couple of other individuals who took part in the hunt manage early on to get a few pieces. But sharing is neither simple nor organized, and to make things much less organized, soon the desperate owner of the meat is surrounded by a mob of chimpanzees who want some. We can imagine that the ultimate goal of each individual in this mob is the same—to get a piece of that tremendously exciting and nutritious food—but the strategies for reaching that goal will be different. Some may be desperately supplicating, possibly hoping that their special status as a friend or a relative or an ally of the possessor may help. Maybe one of the supplicants is a female moving into estrus and wouldn't mind trading with the possessor: a bit of this for a bit of that. It happens. Others may simply hang around and wait to scavenge pieces that fall out during the struggle. And others are actively plotting to pull away something or just preparing to yank and run.

It looks like sheer chaos, but we can recognize that the many strategies divide into two very different responses to the issue of a possessed and highly valued object, and the ultimate result is to create the social phenomenon of ownership. Indeed, one group of chimps has chosen a strategy that confirms ownership, and they, therefore, are prepared to supplicate, trade, or scavenge. For the sake of simplicity, let's just call these chimps *the beggars.* Another set of chimps is actually intent on challenging ownership, and the individuals of this second group are therefore preparing to snatch a piece or grab the entire carcass. Let's call these chimps *the thieves.*

The beggars are often easy to recognize, since they may be making their wishes known to the owner of the meat with gestures and postures virtually identical to those of begging humans: hand held out, palm up, and reaching toward or trying to touch the meat, all done while gazing intently into the face of the individual who holds the meat. The beggars may make pleading or whimpering noises. They can be assertive, persistent, and may even resort to tantrums if they don't get results in any other way. Jane Goodall has described the case of one big male, Goliath, who had caught and killed an infant baboon, and another big male, Mr. Worzle, who was desperately begging for a piece, whimpering and reaching out with his begging hand, following Goliath from one branch to another. Goliath resisted these supplications by pushing away Worzle's hand, rejecting him eleven times before Worzle threw himself backward off the branch they were both on and caught himself on another branch while screaming and striking out wildly at the vegetation around him. Goliath then gazed down at Worzle, that big guy who had just been seized by a compelling tantrum, and tugging powerfully with hands, a foot, and his teeth, Goliath tore the dead prey in half and passed one half over to the frustrated beggar.[4]

The thieves may not be as easy to recognize as the beggars, and maybe that's because the thieves are trying to hide their real intentions. They may act innocuous. They may try to blend in with the beggars. They may gradually be moving closer to the object of their desires. But once they get close enough to the owner and find an opening—*yoink!*—they've reached in and stolen the meat. They run off with it.

This example appropriately suggests that property ownership occurs in response to social conflict. Someone possesses a special object, and others would like to possess it as well, or at least possess pieces of it. The conflict might not be obvious, or conversely, it could seem wildly chaotic. But an examination of the strategies shows us an important dyadic division: two approaches to the problem of someone else who possesses meat, two contrary strategies to fullfill the desire a chimpanzee has for a piece of it. One strategy confirms ownership. The other challenges it.

Yes, these are just chimps. They don't talk or analyze or hold intellectual discussions about the nature and logic of their actions. But they are confronting a problem you and I are familiar with because we live in a world utterly filled with special objects that are possessable—if imperfectly so—and a good deal of our civilization is designed to reinforce the pro-social value that confirms ownership. The physical aspects of civilization confirming ownership include fences and gates, walls and doors, locks and keys. The symbolic bits include deeds and receipts and a dozen or a hundred other versions of written ownership affirmation. The psychosocial machinery includes any number of methods meant to discourage people from violating ownership, ranging from the police and justice system to the moral teachings of our wise elders. As our mothers told us when we were five years old, "When you want something badly, ask for it, buy it, trade something for it. Beg if you must. But don't steal." Why not? Because stealing is wrong. Stealing is bad.

And stealing, so you and I might add, is an antisocial strategy that challenges the existing social order. Chimpanzees don't have a clue about any of these things—and neither do a lot of people. We don't need to. We need only in an emotional sense to appreciate the rules. The rules are there to reinforce pro-social values, and judging from the contents of my morning newspaper, those values do need to be reinforced. What would happen otherwise? What happens when thieves, both simple (robbers) and sophisticated (embezzlers), dominate a human society? Clearly, too much theft will destroy banks, ruin credit, encourage violence, and discourage productive work and charitable sharing. A substantial portion of our system of social cooperation is based upon property ownership, and without the effective protection of ownership, much of that system would just collapse.

In a couple of studies done at Jane Goodall's research site in Gombe National Park, Tanzania, overt attempts to grab and steal meat after a predation succeed only around 15 percent to 20 percent of the time.[5] So thievery of this sort fails more often than it succeeds, perhaps simply because the original owner knows how to defend the prize well, often climbing onto a tree branch, where attacks are physically more difficult to carry out, and locking the meat in a tight, cradling embrace. But what would chimpanzee society look like if the thieves were more successful than they are? Among chimps, as among people, theft discourages anyone from trying to own a special object in the first place. Indeed, if all chimps were thieves, it's fair to assume that no chimp would even bother trying to own meat; and with no one owning meat, no one would have a good reason to hunt prey. Why expend all that energy and risk your life jumping through the trees chasing an agile monkey if past experience tells you that the meat will most likely be taken as soon as you've finished hunting?

Like our own complex system designed to sustain ownership and inhibit stealing, then, the chimpanzee system maintains its form over time through dynamic stasis. Some thievery is both possible and profitable for the individual thief, but too much will unravel the system. The beggars, many of them potential thieves, will yet remain beggars as long as they can reasonably expect to get some meat for their efforts. The owner of the meat, meanwhile, will fight with the thieves and, over time, share food with the beggars, thus promoting a social consensus in favor of ownership.

Another route to ownership, used by certain members of the crow family (corvids: around 120 species, including crows, rooks, ravens, magpies, nutcrackers, and jays), is to outsmart the thieves.

Western scrub jays are large songbirds marked with some bright combination of blue wings, head, and tail; a white throat crossed by a band of blue; a darker brown or gray back; and a lighter gray underside. During mating season, the female builds a nest (twigs, moss, grass, softened on the inside with hair and thin roots) in a tree or bush, from one to ten meters off the ground, while the male stands guard. Then, from early spring to early summer, she'll lay up to a half dozen eggs and incubate them for about sixteen days. Her fledglings will be mature enough to fly away some two to three weeks after hatching. The couple or family may feed together, but outside the breeding season, these jays forage in small groups of unrelated individuals, pursuing their usual diet of berries, grains, nuts, insects, eggs and the hatchlings of other bird species, as well as the occasional bite-size lizard or frog.

Like a number of other corvids, western scrub jays store food for the future. Their storage is simple enough. When they discover a present surplus of good food, they will dig a hole and bury some of it, creating a little cache and thereby, so we might imagine, attempting to make their extra food fast: to own it. The owners remember where they've buried their cache, and they will return at some more hungry time to dig it up and eat it.

This behavior suggests that the jays might have a sense of the future, but do they really? How much of their caching is based on actual anticipation in the sense that you and I anticipate future events? More particularly, are we able to say that they anticipate a "future motivational need"? Western scrub jays can be hand-reared and given a somewhat normal life in captivity, such as within an aviary at Cambridge University in England. There, scientists selected eight of these birds to test whether they could plan for future hunger.

The experiment began with a six-day training session, during which each bird was placed for two hours in one of two different compartments on alternate mornings. In one compartment they could always count on a decent breakfast—pine nuts that had been ground into powder, making them edible but not cacheable. In the other compartment, they would never get breakfast. Other times of the day, they had plenty of food but always under circumstances that didn't allow for ownership through caching. Then, on the final evening following the six training days, the birds were given whole pine nuts, which can be cached. They also were given the opportunity to visit both compartments, the breakfast one and the nonbreakfast one, as well as a connecting area in between the two.

What happened? The birds took their extra pine nuts and cached them inside the nonbreakfast compartment, the place where they learned they would go hungry in the morning. They were anticipating, so the researchers concluded, future hunger.[6]

Following the general structure of this first experiment, the researchers next demonstrated not only that the jays will cache food preferentially in nonfood compartments, but that they will also cache peanuts in areas where they had been given only dog kibble for breakfast, and, conversely, dog kibble in places where they were previously breakfasting only on peanuts. In other words, they appeared to be planning for future food variety.[7]

Western scrub jays are smart birds, as their interesting habit of caching food already suggests. But one day Nicky Clayton, a University of California at Davis scientist interested in animal cognition, discovered a twist in the usual caching routine. Clayton was casually watching scrub jays pick up bits of food left by students on campus when she realized that the birds were not only hiding their extra food in caches but were sometimes later returning to their own hiding places, digging up the food, carrying it elsewhere, and rehiding it. Why would a supposedly intelligent bird waste time and energy rehiding already hidden food? Clayton suspected that the jays recognized that their original hiding places were already known by other jays and might thus be especially vulnerable to theft

Clayton and her colleague and husband, Nathan Emery, designed an experiment using hand-reared scrub jays to test that suspicion. Wax worms were the succulent prize given to captive jays of two sets. The first set of jays received their worms while alone, thus having the opportunity to cache in solitude any surplus worms. The second jays were given worms in the presence of another jay, and they were also allowed to hide their precious prize—but, of course, under the watchful gaze of the other. When both sets of jays were brought to the same place three hours later and allowed to visit their caches in private, the researchers noted that the jays who had originally cached while another bird was watching took the time to dig up their earlier caches and move them to different places. The birds who had originally hidden their worms in private, however, didn't bother rehiding them during this second visit. Rehiding the wax worms, then, seems like a tactic designed to foil potential thieves.[8]

In another experiment, Clayton and Emery demonstrated that a scrub jay would preferentially chose a shady area as the spot to hide a cache, rather than a brightly lit area—but only when another jay was present. Otherwise, without a potential thief around, the owning jays showed no preference for either light or shade.[9] The researchers concluded their jays appeared capable of understanding another bird's visual perspective and had changed their hiding strategy in response.

Not all thieves or potential thieves are alike, however. Scrub jays seem to trust their own mates, at least enough that they are unlikely to rehide a prize that was hidden when only the mate was watching. These birds have also been observed chasing away a potential thief from the cache of a mate.[10] Other than their mates, though, can it be said that scrub jays recognize individual differences among potential thieves? In another experiment, several scrub jays were allowed to hide precious food items in two different sessions, each time under the gaze of another jay. For the second session, however, sometimes the watching jay would be the same individual who had watched the first time, and sometimes not. The researchers found that when the owner jays were returned to the same place for the second session, they distinguished between the watching jays based upon what they were likely to know. If a watcher knew, or was likely to know, where the original cache was—having seen the earlier hiding session—then the owner would now dig it up and move it around multiple times, as if trying to confuse the watcher. But if the watcher could not know where the original cache was—having not seen the earlier session—then the owner would not approach the cache, presumably since doing so would unnecessarily reveal an already successfully hidden prize.[11] Scrub jays, so the researchers concluded, were keeping track not only of the *what*, *when*, and *where* of their food-hiding episodes, but also of the *who*.[12]

All that effort to foil a little burglary! It tells us how significant the problem of theft really is in the social and economic lives of scrub jays and other corvids, many of whom also cache food. Thomas Bugnyar of the University of St. Andrews and Kurt Kotrschal of the Konrad Lorenz Institute, who both study caching and thieving among ravens, find that the thieves appear to spy—lurking in the distance and at the edge of some visual obstruction, such as a rock or tree—on caching ravens, and they typically wait for the owners to leave before conducting a raid.

I have so far described this interesting competition as if the owners and the thieves are different birds. At any one time, they are. But a corvid who has cached food will often try to steal from another bird's cache, so one bird might be both owner and thief at different moments. Indeed, among the more intriguing observations to come from the studies of western scrub jays is that the jays who have had experience as thieves are the more active or astute owners—that is, more likely to re-cache—as if being a thief gives one a better understanding of what's going on in the minds of other thieves.[13]

The same chimps will sometimes own meat, sometimes beg for it, and sometimes steal it, but chimpanzees often specialize. Some are skilled at hunting prey and owning the meat afterward. Others are more likely to place themselves among the beggars. Still others seem to be particularly drawn to become thieves. Satan was among the most active of thieves at Gombe, having been observed over seven years with monkey carcasses that, more than a third of the time, were taken from others.[14] Chimpanzees recognize each other as individuals, so it's entirely possible that Satan was not only a habitual thief but a notorious one. One can imagine that any other chimp, owning a piece of meat and trying hard to maintain that ownership, would stiffen and pay especially careful attention at the approach of Satan.

Scrub jays seem to have a dynamically balanced competition between ownership and thievery, yet it's hard to say that the rules of ownership are sustained by collective rather than individual enforcement. In the chimpanzee system of meat ownership, by contrast, a general dynamic of social enforcement is clear enough. For both owners and beggars, thieves are the problem; and, for siding with the owners, the beggars are rewarded by getting their share of the food. Sharing food might sound like altruism or simple generosity, chimps running a soup kitchen, but the reality may be less soup kitchen and more marketplace. A marketplace means nongenerous chimps have a series of mutually beneficial exchanges that might be prompted by gestures, vocalizations, physical actions, or may just silently be understood as part of the background social knowledge. What kind of exchanges? Sex for meat. Reduced tension for meat. Simple peace and being left alone for meat. Past friendships confirmed or future political alliances made more likely for meat.

Ownership makes that marketplace possible.

For people, the three kinds of potential property—object, real estate, or another person—are marked by three levels of difficulty in possession and therefore in ownership. Owning an object is comparatively simple. You might, in the style of chimpanzees, grasp it tightly and enlist a surrounding crowd of allies who affirm your right to ownership, or at least don't challenge it. You might, in the way of scrub jays, try hiding your object in a safe place and making sure no potential burglars are watching as you do. A piece of real estate, though, is too big to grasp or hide, so owning real estate requires possession through some kind of expanded defense. And trying to own another human being, either as an utterly subordinate spouse or as a slave, implies a set of radically more difficult problems. The spouse might get tired of being subordinate and leave in the middle of the night. The slave might slit your throat.

Animals have those same three kinds of potential property. Owned objects, as we have seen, are mainly important food items. Real estate? Many animals claim ownership of real estate, and they sometimes do it collectively, with particular social groups of a species cooperating to keep their territory safe and separate from that of other social groups of the same species. Individual territoriality is commonly an aspect of the *lek* system. A lek is an arena where the breeding adults of a species gather to court and mate, and often in a lek the males will take over their own miniterritories and try to attract females. This system appears among several kinds of African antelope and a number of bird species, such as the sage grouse of the western plains in North America.

A lek for sage grouse might cover a hectare or more with a smaller area at the center where mating actually takes place. At breeding time, a few hundred males coverge on the lek and stake out individual bits of real estate, each one only about ten to a hundred square meters in size. A few hundred females also arrive at the lek, where they are eagerly courted by the males. The males puff out their chests and strut. They raise their white neck feathers and elevate their yellow eyebrow combs. They stretch their wings exaggeratedly forward and back, revealing a pair of olive-yellow skin patches on their chests while producing a swishing sound, as feathers rub against feathers, then a snapping sound as air in a chest sac is rapidly compressed and released into the skin patches. The males also coo, and so their dramatic visual display is accompanied by an auditory one that has been summarized as *swish-swish-coo-oo-poink*. All the adult males gathered at the lek are apparently able to perform this wonderful dance with some facility, but only those holding territory that overlaps the part of the lek where mating occurs are likely to win a female's full response. Indeed, one tenth of the males get three quarters of the sex, which means that some pieces of real estate are a lot more valuable than others.[15]

You and I are used to marking our real estate symbolically and visually, with maps and deeds, flags and fences, while many animals try to identify their real estate with scents, sounds, gestures, or actions. This is so common we don't think twice about our own favorite canine's strange habit of cocking a leg and depositing scent-waifs along the sidewalk and around the block. But what kind of animal tries to own another?

Slavery—and its rough equivalent on the domestic scene, the utterly subordinated spouse—seems at first glance to be one of those disturbingly unique marks of our own kind. Because owning a person is difficult to do, it could be, so we tell ourselves, that the slavery we know from our history books, and occasionally from our daily newspapers, requires a certain level of technology to enforce long-term control: chains, shackles, and the like. Animals don't have that technology. Do they have any kind of conspecific ownership that might be reminiscent of human slavery or its domestic equivalent?

A kind of slavery has been observed among some ants, where the soldiers of one colony raid another colony of the same or a related species, stealing pupae and bringing them back home to raise as workers.[16] But this, at least when it involves individuals of a related species, is more like domestication rather than enslaving one's own kind. Conspecific ownership—ownership of individuals from one's own species—is perhaps more comparable to the social arrangements of some polygamous mammals, where one sex seems to exercise signficant control over the other. This usually means males controlling females, and it seems to be one consequence of sexual dimorphism. Where sexual dimorphism is extreme—males twice as large as females, as with elephants, gorillas, orangutans, and others—the males may use that extra size and power not merely to compete with other males for mating but also to control the females.

Consider, for instance, hamadryas baboons. As an aspect of their normal social and sexual lives, the males acquire harems of females and herd them somewhat in the style of a sheepdogs herding sheep—but big sheepdogs with daggerlike canines. We can imagine the females as passive victims of their species' social system, but, of course, this is the only system any of them know or can imagine. A female might try to escape her own rough and overbearing male, but she would merely escape to another herding male and into another harem. Indeed, a smart female might actually choose a bigger and stronger and more domineering male than the one she already has, since bigger and stronger and more controlling are the marks of an evolutionary winner: a mate more likely than average to give her evolutionary winners as offspring. Females, then, are also participants in the system.[17]

The rules of ownership, so Melville first insists in "Fast-Fish and Loose-Fish," are simple, and they apply not merely to a dead whale on the high seas but to anything else that has the potential to become property: object, real estate, or person. But he expands on this idea with a rhetoric that turns increasingly complex and fanciful. Object: "What to the rapacious landlord is the widow's last mite but a Fast-Fish?" Real Estate: "What was America in 1492 but a Loose-Fish, in which Columbus struck the Spanish standard by way of waifing it for his royal master and mistress?" Person: "What are the sinews and souls of Russian serfs and Republican slaves but Fast-Fish, whereof possession is the whole of the law?"

This expanded view of property is confirmed by the tenth commandment, which prohibits even desiring certain kinds of property belonging to one's neighbor: "You shall not covet your neighbor's house; you shall not covet your neighbor's wife, or his manservant, or his maidservant, or his ox, or his ass, or anything that is your neighbor's."

Both statements recognize the importance of ownership and the multiplicity of what constitutes owned property—but how different they are. Melville's words are dripping with irony. The biblical commandment, authored by God or Moses or some comparable human authority, includes no sense of irony whatsoever. Not a whiff. Not even the sense that irony might exist. Irony, after all, is not appropriate when you're God, cloaked in thunder and lightning, or Moses returning from the top of a mountain to present a new set of laws to a potentially rebellious people. It's easy to see why God or Moses would be incapable of irony, but why would Melville appear so thoroughly aligned to the ironic stance? Why, in other words, should Melville express such an emphatic ambivalence about the fast-fish and loose-fish rules, which at first seemed to be so appealingly simple and straightforward?

The answer is clear enough. Melville was acutely aware that other moral rules or ideas or inclinations can conflict with or supersede certain aspects of the rules of ownership, and the reader of *Moby-Dick* is made to understand that. The reader of the Ten Commandments in Exodus 20, by contrast, moves on to chapter 21, which presents the several ordinances detailing how, precisely, one should treat one's own slaves.

CHAPTER 9

Communication

"Jonah did the Almighty's bidding. And what was that, shipmates? To preach the Truth to the face of Falsehood!"

—Herman Melville, *Moby-Dick*[1]

My dogs can be trusted when good food—fancy crackers, three kinds of cheese, black olives, a little pâté—is placed at nose and mouth level. Let's picture, here, a festive situation with a number of interesting and animated guests surrounding a plate of hors d'oeuvres placed on the coffee table in my living room. Yes, Smoke and Spike can be trusted. They may weave through the forest of human legs to investigate such treats with their giant sniffers, but they always do so with discretion, demonstrating that a nose four inches away from the plate remains on the proper side of the line, whereas a nose two inches away has moved to the improper side. They don't ordinarily cross that imaginary line, and they will never just snatch some tasty item and dash off.

Never, that is, if any responsible person is present and watching. But try moving everyone to the kitchen for more than, say, three minutes. You then come back to find a living room from which all the cheese has disappeared and the cracker supply looks suspiciously diminished. A theft has been committed, but the dogs are quietly lingering elsewhere with, so it seems to me, bland expressions of complete innocence on their faces. Or is that a hangdog expression of guilt I see?

Whom do they think they're fooling?

But serious deceit can cause serious damage. Say, for instance, you've gone out on a dark evening all by yourself, looking for sex, just a little normal, ordinary sex, and you're quietly cruising through the flashing-light district. Since you're a male firefly of the *Photinus consanguineus* species, you are sending out your own distinctive signal of availability: a double pulse of light, one separated from the other by two seconds, followed by a delay of four to seven seconds, then another double pulse. Now you're just cruising about and looking for the responding signal of an available *Photinus consanguineus* female. About a second after the end of your double pulse, she should respond with a single pulse. . . . Oh! Over there! You see the responding signal. That signal quickly becomes a beacon, which you follow right to its enticing source until . . . arrgh . . . too late! You've been tricked by a large and predatory female *Photuris* firefly, craftily sending out a deceitful imitation of the female *Photinus consanguineus* signal, and now you're dead.[2]

That's deceit in the extreme, and it demonstrates at least two things. First, it reminds us of the dyadic nature of truth and falsehood, or what I might call, in this case, *honest communication* versus *deceptive communication.* I describe such a pair as dyadic because each half is locked in a dependent embrace with the other. The deceptive signal of the *Photuris* predator firefly would, after all, be useless without the existence of an honest signal ordinarily sent by willing *Photinus consanguineus* females. Conversely, it's illogical to describe the signal of a willing *Photinus consanguineus* female as *honest* unless it stands in contrast with the potential of some dishonest signal somewhere. Honesty and deceit make sense only in tandem. Recognizing dyads is essential for understanding moral behavior. It can also be useful for those of us looking to find moral behavior in animals, since one half of the dyad—deceit, for instance—may be much easier to identify than its partner.

The case of the predatory *Photuris* female fatally fleecing the amorous *Photinus consanguineus* male demonstrates a second thing about honesty and deceit: Honesty serves the interests of both parties, whereas deceit serves only one participant. In our human world this interesting difference means that we recognize honesty as a pro-social and often a moral option, whereas deceit is likely to be seen as selfish, antisocial, and—if serious enough—immoral.

Domestic roosters sometimes make food calls in the absence of food as a clever if deceitful way to attract unwary hens.[3] Frogs deceive one another, as do crabs. So do supposedly monogamous birds, with DNA assessments showing an average of 13 percent nonmonogamous matings.[4] Well, the universe is filled with communications honest and deceitful, which becomes only too evident the minute we turn the looking glass back onto our own social world. Bella DePaulo, a psychologist at the University of California at Santa Barbara, organized a survey in which seventy-seven students and seventy nonstudents agreed to keep a week-long diary of all their social interactions and any lies told during those interactions. The study's 147 participants described telling some 1,535 lies, with an average of one lie for every five social interactions. Only one of the college students claimed not to have told a lie during the week, while six of the seventy nonstudents also reported being entirely honest.[5] One can only wonder how much longer it would take before all participants could honestly report having lied. Or were the handful claiming total honesty during the week lying about that?

We rationalize some of our prevarications by describing them as necessary acts of kindness, as in *Did you really expect me to tell him he's a boring jerk?* And indeed many of the lies in DePaulo's study appear to have been uttered to avoid hurting someone by stating a painful truth. This sort of lie appears to violate one moral rule (against deceit) to favor a second moral rule (against inflicting unnecessary pain). DePaulo recognizes the existence of such lies, which she describes as being *other-oriented*, but even if we eliminate from her study all of those lies rationalized by an appeal to kindness, the data still support a rough average of one lie per day per person.

Most everyday lies are likely to cause little damage. Yet even if we classify minor lies as lapses in manners rather than in morals, we may still conclude that most people tell genuinely serious or harmful lies from time to time. These are deceits meant to advance one's own welfare, hide one's secret activities, harm one's adversaries. Among humans, at least, we understand clearly that a moral principle is involved in serious lies. Philosopher Sissela Bok states the case emphatically by comparing deceit with violence, describing both as "two forms of deliberate assault on human beings. Both can coerce people into acting against their will."[6]

Whereas a truth-teller may enjoy the fruits of that truth, and a liar might enjoy the results of some lie, no one—neither truth-teller nor liar—likes being lied to. Truth-telling is socially welcomed, while telling a falsehood is only privately perpetrated. Thus, telling the truth will be found on the list of moral virtues, as good and pro-social, while lying belongs on the list of moral vices, as bad or antisocial. Of course, we recognize a tremendous individual variability in the inclination to deceive. Some people lie more than others. But in general, we can conclude that the human capacity for antisocial deceit is almost bottomless and endless, while, at the same time, the ninth commandment as written in Exodus 20 weighs in with a powerful if limited attempt at pro-social counterbalance: "You shall not bear false witness against your neighbor."

The ninth commandment might be read as a specialized prohibition having to do with formal testimony in a court of law, but everyone also recognizes this more general rule: Preach truth to the face of falsehood.

As I expect you've already concluded, the story of the predatory *Photuris* and a preyed-upon *Photinus consanguineus* may evoke the honesty-and-deceit dyad appearing in nature, but it is not a legitimate example of animal morality or immorality. I earlier described morality as ordinarily restricted to interactions among individuals of the same species. For a member of one species to deceive an individual of another is no different from a person catching a fish with a clever lure. No different from one creature using camouflage or stealth or any other piece of ordinary trickery to prey on or to avoid becoming prey of another creature of another kind. Other species are usually fair game. Moral rules do not usually cross the species barrier. That's the effect of Darwinian narcissism applied to morality.

No, if we're going to search for moral behavior in the realm of communication, we should look at communication between individuals of the same species. When we do so, we find that even at the simplest levels, the use of any forthright communication can provoke a deceitful alternative. Take the Augrabies flat lizards, for instance.

These remarkable reptiles live in world-class abundance around cliffs that line the banks of the Orange River of Augrabies Falls National Park in South Africa. They're dramatic little creatures. They'll zip headlong down a hundred-meter-high vertical rock-face, and in pursuit of insects, they do backflips. They're also adapted to living in the rock crevices so common in this place. The females lay their two yearly eggs down in a crevice during the height of summer; and both sexes sleep in those same inaccessible recesses, their tails wrapped around their bodies for protection from both cold and predators. The tails are replaceable, so should a predator—a bird, for example—grab a flat lizard by the tail, the reptile can simply drop it and skitter off to safety. A lizard is given only one life but many tails.

Their social world converges on the source of their most important food: blackflies. These nutritious insects rise in swarming plumes at particular spots along the river's edge; and from the lizards' perspective, of course, those spots are the most desirable plots of real estate. The better territories are taken by the bigger and more vigorous adult males, each competing with the other adult males to claim and maintain that valuable real estate. Naturally, possessing superior real estate not only translates into nutritional benefits; it also means reproductive benefits. While the resident male chases away the other males, the observant females will eagerly crowd in, always interested in those places where the blackfly plumes abound.

Good food equals good territory. Vigorous males get better territory. Females choose better territories and the vigorous males who own them. Nothing surprising here. This is an ordinary, predictable, and stable social arrangement.

Adding to the stability is the fact that the males, once they reach full adulthood, begin to take on their coat of many colors: a brilliant wash of yellow, orange, green, and various shades of blue. The adult males, already significantly larger than the adult females, thereby stand out with great vividness—like a flag. We should think of their vivid coloration as a visual communication to the females: Here is a powerful male who is likely to have claimed good territory. Since the larger and more aggressive males—that is, the better fighters—actually have brighter coloring, their colors also allow the males to identify who is more likely to win a fight. Thus, the coloration also reduces actual fighting. The critical areas are located on the abdomen in patches of bright orange or yellow, and the males use their abdominal skin patches like badges of rank. Two males square off, flash their abdomen patches at each other, and if one can show he clearly outranks the other, that settles it. No need to fight.

This is an example of honest communication, and it provides some obvious benefits, but it also comes at a real cost. Overhead, high above the rocky ledges and cliffs and all those wriggling, squabbling, mating, blackfly-eating reptiles, there turns and turns a predatory bird, the kestrel, just looking for a good lizard to eat. You would think that during the millions of lizard generations this sort of thing has been going on, natural selection would have produced increasingly hard-to-see lizards, ones camouflaged to look just like the rough and splintered rocks on which they congregate. In fact, that very thing has happened. The females and the immatures of both sexes are colored a dull brown with a pattern of dark and light stripes on their backs that resembles the pattern of little rock crevices. That makes perfect sense. But we can also imagine that the adult males have abandoned that sense in favor of another sense, one that strongly values the communication they give to both other males and to females.

A good bright skin makes other males back off, and it makes females move in. That's its value. The cost of this coat of many colors, this flag or visual communication, is what keeps it honest, since any male lizards wrapped in such a vivid skin are defying the predatory intent of that overhead kestrel. Vividness is an honest communication of vigor, and we can see that the system has developed its own natural barrier against deceit.

Deceit happens anyhow.

Since the juveniles of both sexes maintain the same dull, rocklike coloring, the maturing of males is accompanied by a gradual development of bright colors. Such young adults are still not big or self-confident enough to hold their own, and they are thus vulnerable to attack from fully matured males. Since they can't seriously compete for territory, they have no good way to attract females. Too bad for them. However, a portion of these young males reach sexual maturity before they acquire the full coloring that identifies a mature male. They still look like females, and they use this temporary and circumstantial disguise—a male in female clothing, as it were—to their advantage. While the big, brilliant, vigorous male is standing in the middle of his glorious real estate, observing the lovely plume of swarming blackflies and surveying all the attractive females assembled there in huge numbers, he is also watching out for any intruder males. Any intruder he sees, he will do his best to chase and bite, chase and bite, until the intruder retreats. The younger male who still looks like a female, however, may be able to avoid such an unpleasant confrontation. He may simply be able to settle himself inconspicuously in the midst of a seething crowd of real females. If he's lucky, he'll now have plenty of opportunities to mate right inside the territory of the guarding resident male.

The resident male may be entirely fooled by such a visual deceit. But there's the olfactory problem, caused by pheromones. Male lizards smell different from females, and the resident male is likely to go around sniffing, which he does by licking with his tongue. This means that the young male who looks like a female but still smells like a male has to remain on guard. When he sees the resident male approaching, he will move away, keeping always a few steps ahead of the probing tongue, careful always to avoid having his big deceit uncovered by a flickering lick.[7]

It's an odd story, isn't it? But does this strange case of animal deceit have any resemblance whatsoever to our own human experience of lying?

In some important ways, yes, I believe it does. For one thing, since all deceits challenge a social norm of honesty, all deceits have in common the possibility of rule enforcement. Creating trouble. Getting caught. Being punished. It is not hard to imagine that the deceitful male Augrabies flat lizards are motivated to avoid the physically and socially powerful resident male's probing tongue through a fear of being caught and punished. Among people, the fear of getting caught in a lie, either because of the likely shame or the likely punishment, makes our heart race, our respiration spike, and our sweat glands work overtime—all physiological changes that the standard polygraph test is designed to assess. Indeed, such events associated with general anxiety are predictable enough that a new generation of technologies is being developed to refine lie detection. Thermal-imaging devices, for example, may soon enough be deployed in airport security barriers to detect the minor temperature elevations (as little as half a degree Fahrenheit) that result from an increased blood flow into tiny capillaries around the liar's eyes.[8]

For humans, though, lying not only increases our general anxiety, but it requires a significantly engaged intellect. Yes, truth-telling also requires an engaged intelligence, since it considers someone else's knowledge and ability to understand. But real lying in the way of humans requires more. In truth-telling, a person merely expresses what he or she has in mind. In lying, a person retains that one true idea while simultaneously conjuring a cover story presenting a second, false idea. Among humans, moreover, lying is not just a matter of creating such a misleading communication in spoken language. Since people communicate so much additionally through facial expressions and gestures, lying also means misrepresenting oneself facially and gesturally. Liars deliberately create a mask: a face of falsehood that, given the complexity of our normal facial communications, is in itself a remarkable achievement.[9]

The immoral act of lying, human-style lying, could be the most intellectually demanding of the vices, since to perpetrate a lie requires real craft. Yes, deceit happens commonly in the animal kingdom, as we have seen. But lying as humans practice it requires fooling other humans, and since other humans are exceptionally smart and quick about not being fooled, the successful liar has to be even cleverer and quicker. Let us think of human-style deceit as a mind game in which the deceiver must be aware that the victim has a mind or a certain degree of mental awareness. For the lie to succeed, the liar must make sure the victim's awareness remains at a state that is different from the liar's own awareness. Cognition theorists sometimes refer to that sort of high-level lie as *intentional deceit*, and they say that it shows the intelligence necessary to possess a *theory of mind.*[10] Nothing in my story of life among the Augrabies flat lizards suggests such a high degree of intelligent awareness or a theory of mind.

Intentional deceit implying theory of mind among animals would mean something more on the order of what the fictional whale Moby Dick does. Some of the sailors know about the legendary whale based on his appearance, but what really marks this big guy from the others is his crafty deceptiveness. Moby Dick is smart enough to anticipate what the whalers are planning and simultaneously to conceal his own real plans in an elaborate drama of falsehood, with his favorite trick being to lead a team of naive whalers on a chase. The whalers, distracted by the sight of so much valuable flesh and blubber and oil wrapped up in one skin, row eagerly, moving farther and farther away from the safety of their mother ship as the great creature before them swims with a convincing counterfeit of panic. At the right moment, however, having lured them far enough onto the grasping hands of a restless sea, the beast will turn about and chase his chasers, finally to smash their little pursuit boats to pieces or to harass them maliciously until they've clambered in genuine panic right back aboard the mother ship. That is intentional deceit, and it implies an awareness of someone else's mental state: the whale knowing and exploiting the perceptions and expectations of the whalers.

But aside from this imaginary whale from imaginative literature, where else among animals can we find evidence of such a sophisticated awareness—and how will we know it when we see it? The answer to this two-part question is not as simple as one might hope. When I wondered at the start of this chapter whether Smoke and Spike, having stolen hors d'oeuvres, were trying to hide their guilt behind expressions of innocence, I was really wondering about dog intelligence. I think dogs are probably capable of attempting deceit, but what about feeling guilt? Maybe you'll say dogs just aren't brainy enough to feel guilt, that they harbor only a reflexive fear, having learned from past experience that a big fuss usually follows an hors d'oeuvres theft. That's anticipatory fear. But is it possible that what people like to consider *guilt* is really nothing more than the same kind of anticipatory fear with some overlying complexity: an embedded splinter of dread based on specific acts and learned responses?

If you have a ball-playing dog and want to think about a dog's ability to read what's going on in your mind, just watch. Your dog will bring the ball up to your feet, drop it, assume the play position, then move a gaze intently from the ball to your eyes and back to the ball again. *When are you going to see the ball and recognize my eagerness to play with it?* the canine seems to be asking, while tending to the quality and direction of your gaze for an answer.[11] Dogs seem to follow gaze direction to gain information about others' perceptions and intentions, and they appear to use eye contact to initiate and emphasize communication.

So, according to an account recorded in *Baboon Metaphysics* (2007), by Dorothy Cheney and Robert Seyfarth, do some monkeys. Namibian conservationist Conrad Brain, sitting on a heap of rocks in a desert canyon and watching a troop of baboons, noted the one juvenile female who approached him and looked into his eyes. He returned the gaze. Her eyes then turned down in the direction of a rock crevice at the man's feet. She gave an alarm bark and looked back into his eyes. She gave another alarm bark, whereupon he looked down to where the baboon had just been looking to discover the object of her alarm, which was a spitting cobra. Brain, backing away from the cobra, retained the powerful sense that the baboon had acted deliberately, intentionally making this honest communication about a serious danger.[12]

But had she really? We can't know. We can know that dogs, monkeys, and apes share with humans the neurological ability to follow gaze direction and, in some degree, to interpret it. The superior temporal sulcus (STS) is a part of the brain distinctly attuned to gaze direction, and a mutual gaze—as shared, for example, by a baboon and a conservationist—provokes an increased response in the STS for both individuals of both species. Moreover, neurons in a monkey's STS light up in functional magnetic resonance imaging (fMRI) scans whenever that individual observes another monkey looking at his or her hand during a goal-directed movement, so it could be that this area of the brain is also important for representing what someone else is seeing and intending. Monkeys also share in common with apes and humans the so-called *mirror neurons*, located in the inferior parietal lobule (IPL) of the brain, which become active both when the individual is performing a particular action and when he or she observes someone else doing it. Neurological research, then, suggests that monkeys and apes can appreciate the connection between seeing and knowing, and thus possess an awareness of the awareness of others.

So the deceit of monkeys provides a convenient means for measuring the intelligence of monkeys. But the real masters of deceit in the nonhuman universe will probably be found not among monkeys at all, but rather among those animals who are even closer to us on the evolutionary tree, apes—and of the apes, we might usefully consider those who are both the most studied and the most popularly known for their wily ways.

Chimpanzee deceits, like all deceits, rely upon a background expectation of truth, and they appear in two different varieties. In deceits of omission, true information has been suppressed, thereby creating a false impression. In deceits of commission, false information has been expressed as if it were true.

Deceits of omission are commonplace. In his book *Chimpanzee Politics* (1982), primatologist Frans de Waal describes several instances in which chimpanzees at the Arnhem Zoo in Holland suppressed true information to pursue a deception. In one case, a young male named Dandy was excitedly courting a female by, in the usual way, showing her his inspired erection, when suddenly an older and higher-ranking male appeared on the scene. Big trouble. Worse, the young male's forbidden interest in the female was being communicated honestly by something he seemed to have little control over, an erect penis—which he quickly covered with his hands, obviously hoping thereby to hide his signal of interest from the alpha male.[13] But not only males try to get away with sex that is forbidden by the socially powerful. De Waal writes of Orr, an adolescent female at Arnhem, who would scream while she was having sex. During surreptitious copulations with younger males, however, her screams sometimes caught the attention of the alpha, who would do his mighty best to interrupt the couple. Eventually, Orr learned to suppress her vocalizations when mating with lower-ranking males, while she continued screaming whenever she mated with the alpha.[14]

Another story from Arnhem. The younger male Luit was beginning to challenge the leadership of the alpha, Yeroen. Luit was still smaller and probably less powerful than Yeroen, so his challenges required not only plenty of politicking—developing that important coterie of supportive friends and relatives—but also a good deal of nerve. Where nerve failed, as Luit must have understood, the appearance of nerve would have to do.

After one of his periodic conflicts with Yeroen, for instance, Luit was observed turned away and three times using his fingers to draw his lips together, in an apparently deliberate attempt to remove the grin of abject fear that had overtaken his face.[15]

Competition often motivates individuals to deceive in order to get what they want, whether it's sex, power, or food. Once at Arnhem, the chimpanzees all observed the arrival of a box of grapefruit. While they were locked in their sleeping quarters, however, de Waal brought the box out into the chimp colony's public area and buried the grapefruit in sand. He left a small portion of each grapefruit still uncovered by the sand, just enough for an observant chimpanzee to notice. After the fruit had been buried, de Waal walked past the chimps with the empty box, so when they were released from their night cages, they raced off in search of the hidden fruit. Several rushed and scrambled right past the place where the special treats had been buried, but none paused to examine that area carefully. Later on that day, however, as the Arnhem chimps were relaxing during their regular afternoon siesta, Dandy, a young male who had been among the group that earlier rushed past the buried grapefruit, now quietly raised himself from his relaxed sprawl, strolled over to where the grapefruit had been hidden, and, away from the gaze of his relaxing fellows, dug out the fruit and consumed it at his leisure.[16]

Jane Goodall recalls several similar incidents among the chimpanzees of Gombe National Park, in Tanzania. During her early years there, Goodall began provisioning the chimps with bananas, but at first she had no good method for doling out the fruit methodically. When she and other researchers brought out the day's bananas, any nearby chimps would rush and grab, which meant that the smaller and younger chimps were sometimes frustrated. One time, for instance, nine-year-old Figan was unable to get even a single banana, so he remained sadly at the provisioning site after the other apes had eaten and wandered off. A researcher then handed Figan several bananas. Tremendously excited by this good fortune, the young chimp eagerly responded with loud food barks—a normal and honest response to food. That honest communication worked only too well, though, and soon several others had returned to the site and taken all but one of the bananas out of Figan's hands. Too bad, so sad. But he never made that mistake again. The next day, Figan stayed behind once more, and was once more given his own armful of bananas. The researchers could hear "faint, choking sounds in his throat," evidence of Figan's struggle to suppress the excited food communication. He succeeded and was soon sitting there in near silence, peacefully consuming his bananas. Dishonesty has its rewards.[17]

But chimpanzees not only are capable of dishonesty through suppressing an honest communication; they are also adept at expressing false communications. Feigning boredom, for instance.

When he was still an adolescent, Figan one day acquired the carcass of a colobus monkey. Goodall watched him sitting in a tree, slowly pulling apart the carcass and smackingly consuming the pieces of meat, once in a while letting his baby sister pluck out bits to chew on. While both siblings were in the tree with the carcass, Goodall saw their mother, Flo, also in the tree, on a lower perch, showing absolutely no interest in the meat. Her apparent lack of interest seemed strange since Flo was known to love meat, but now, with her two offspring up there eating some, Flo was not even looking in their direction. Why? After a few minutes, Flo moved slightly closer to Figan. Still, she didn't look his way or show any other sign of interest. After a while, she climbed up the tree even closer, this time within reaching distance, but instead of looking at that boring old piece of meat, she acted as if she were most interested in grooming herself. Seven minutes passed, then Flo, in a flash, grabbed for the drooping tail of the dead monkey. Figan must, however, already have understood this attempted deceit and theft, since even more quickly he yanked the prize out of his mother's reach.

Flo returned to showing total lack of interest in the meat, gazing idly about. She continued occasionally grooming herself. A few minutes later, having once again crept closely enough to the object of her true interest, she made another quick grab at the meat. Again, Figan was too quick.

So it went. For a full hour, Goodall watched and took notes on this drama between a mother and her maturing son. He with the meat, the informed wariness, and the quick reflexes. She with the desire for meat, the slow creeping, and the feigned boredom.[18]

You may have noticed that feigning boredom, a dishonest communication, is not so far removed from merely suppressing the honest communication of interest, but the chimpanzees at Gombe and Arnhem have shown themselves to be additionally skilled at dishonestly distracting one another or dishonestly diverting someone else's attention, or of faking for dishonest purposes an interest, an injury,[19] or an emotion.[20]

In a classic laboratory experiment, four young chimpanzees, one male (Bert) and three females (Jessie, Luvie, and Sadie), were introduced to games involving hidden food treats: candy and cookies, fruits and berries. Each chimp played the game individually, beginning in a cage inside a larger room. Someone would bring a treat into the room and ostentatiously place it underneath one of two containers. The chimps always saw where the food was hidden, but their cages restrained them from reaching it. Fortunately for them, more people would come into the room who might be able to help out. None of them had advance knowledge of where the food was hidden, and they had only one minute and one chance to find it. The people had, moreover, been instructed to dramatize the role of either a cooperative person or a selfish one. A cooperative person discovering the right container with the treat during that minute would hand it over to the chimp. If the selfish person found the treat, however, he would eat it himself, leaving nothing for the chimp.

The four young animals responded positively to the cooperative person, almost immediately learning to communicate to him which container held the treat—pointing, usually, with their gaze or body posture, or a hand or foot. The selfish person presented a more difficult problem. The oldest chimp, Sadie, learned first, and the youngest, Jessie, figured it out last. But eventually all four withheld information from the selfish person about where the food was hidden. By the end of the experiment, two of the four chimps had actively begun giving the selfish person false information, pointing in their various ways at the empty container instead of the one that actually contained a treat.[21]

People who have formed relationships with captive chimpanzees quickly discover that these apes will usually communicate honestly and simply, but that they will also regularly try a variety of deceitful tricks to outwit their human companions.

Roger Fouts began his psychology graduate work in 1967 at the University of Nevada by working as a babysitter and teacher to Washoe, the first chimpanzee to acquire a facility with human language—in this case, American Sign Language, the gestural language of the deaf in North America. The two-year-old animal, dressed in diapers and sweatshirt, seemed in many ways like a two-year-old child, only far more athletic. Washoe already had a working vocabulary of about two dozen gestural signs, including nouns, pronouns, verbs, and modifiers, when Fouts arrived; and she was using those signs spontaneously in meaningful combinations to comment about the environment around her (LISTEN DOG), claim possession of her doll (BABY MINE), and give instructions (YOU ME GO

OUT HURRY). She would soon learn to brush her teeth, linger over and comment on the pictures in a book or magazine, use a toilet, sew random stitches with needle and thread, loosen screws with a screwdriver, and so on.

Washoe's living and sleeping quarters were inside a trailer parked in the fenced backyard of the two scientists who had begun this remarkable experiment, Allen and Beatrix Gardner. Of course, everything in that trailer—including the refrigerator and all the cabinets and other storage compartments—was kept locked, since the active little ape could climb anywhere and, like any normal human child, was inclined to get into trouble. The entrance door to the trailer was left unlocked only when Washoe was outside, playing in the yard with appropriate supervision.

In his memoir and assessment of the work with Washoe, *Next of Kin* (1997), Fouts recalls a day in the summer of 1968, soon after the chimp had turned three, when he and she had left the trailer and gone into the yard to enjoy the warm weather. Fouts had seated himself on the doorstep of the trailer and was writing notes in a logbook, while Washoe entertained herself by climbing into a tree in the yard. A few minutes passed, and then Fouts noticed that the chimp had left the tree and was now in the rock garden, watching with great interest some small drama taking place in the shadow of some rocks. What was she looking at so intently? Fouts became curious enough that he stood up from his seat on the trailer steps, wandered over beyond the tree to the rock garden, and began looking down into the shadow beneath the rocks to see just what this chimp found so interesting. He saw nothing. Washoe remained there, however, looking very, very intently at something. Fouts finally gave up trying to see what Washoe found so interesting and sat down at the edge of the rock garden.

Once he had settled down there, however, Washoe moved back to the tree and climbed back to the top. All was well. Fouts returned to writing in the logbook . . . looking up at just the moment the chimp had dropped like a stone right out of the tree and was racing to the trailer. By the time Fouts realized what was going on, it was too late. Washoe had already burst through the trailer door, had found the very cabinet Fouts had forgotten to lock that morning, and was now coming out with a cold soft drink tucked under one arm. With the arm thus engaged, she was unable to "run on all fours," as Fouts recalls, "so she was staggering toward the tree on two feet like a drunken sailor." Within a few seconds, she was up in the tree, safely out of Fouts's reach and thus able to finish the stolen drink at her leisure. "The whole episode amazed me," Fouts writes. "Washoe must have noticed that the cupboard was unlocked during breakfast, suppressed her natural impulse to raid it when my back was turned, and instead devised this plan for distracting me long enough to gain access to the trailer by herself and give herself the opportunity to drink the soda. This was a level of planning and deception beyond anything I thought her capable of."[22]

After he completed his graduate work, Fouts moved to Oklahoma, where, in a project associated with the University of Oklahoma, he continued working with Washoe while also teaching sign language to a number of other chimpanzees. One was a young female named Lucy, who was also being experimentally raised as if she were a human child in the home of clinical psychologist Maurice Temerlin and his family. Lucy was by all accounts a remarkable animal, emotionally attached to her human family, lovingly overbearing with her pet kitten, and a quick study, having mastered a vocabulary of seventy-five signs within her first two years of language school. That was enough of a vocabulary for Lucy to produce what Fouts describes as the world's first language-based lie from a chimp.

Like Washoe, Lucy had learned to use the toilet appropriately at an early age. Occasionally, though, she would have an unfortunate accident, such as the one that left a mess in the Temerlins' living room one day.

Fouts was there and used sign language to discuss the problem with Lucy. WHAT THAT? he asked the chimp, using the pair of signs and pointing to the offending object.

Lucy repeated Fouts's signs back to him, as if she had no idea what he was talking about: WHAT THAT?

Fouts challenged her: YOU KNOW. WHAT THAT?

Lucy correctly responded with the sign for feces: DIRTY DIRTY.

Fouts pressed on: WHOSE DIRTY DIRTY?

Lucy named a graduate student who had recently been there: SUE.

Fouts: IT NOT SUE. WHOSE THAT?

Lucy next identified Fouts as the guilty party: ROGER.

Fouts: NO! NOT MINE. WHOSE?

Lucy finally admitted the truth: LUCY DIRTY DIRTY. SORRY LUCY.[23]

It's nice that Lucy apologized at the end, maybe—but was she apologizing about the original forbidden act or the lie that followed it?

We often use deceit to cover some other moral failure, but one distinctive thing about human lying is that we understand that the lie itself is wrong. Lies confuse others, inhibit others, hurt others, so we understand that lies are antisocial and belong on the list of antisocial moral vices. Other people usually expect the truth, after all, and when we violate their expectation, we've done a wrong thing that is sometimes worse than the bad behavior we were trying to conceal in the first place. That, at least, is the wisdom imparted by our mothers when they said, "I don't mind that you took the cookie from the cookie jar half as much as I mind that you lied to me about it."

Chimpanzees often deceive each other, that much is clear, but is it also true that chimpanzees, like our own mothers, dislike being deceived? No one has thought to ask this question, and it would be a hard one to study experimentally, so I will conclude this chapter with a simple story on the subject as recollected by Frans X. Plooij, a primatologist at the International Research Institute on Infant Studies in the Netherlands.[24]

Like several prominent primatologists of his generation, Plooij was first trained in field research at Jane Goodall's research site in Tanzania, where he studied the group of chimpanzees already made famous by Goodall's early work. As Plooij recalls, at Gombe all the researchers were forbidden from interacting in any way with the chimpanzees. The reasons for that rule were obvious. Interaction could affect the research results, so it was bad science. Interaction could also endanger the chimps, who are capable of contracting virtually every infectious human disease. Finally, it could endanger the people, both through disease transmission and also through plain physical damage, especially should the chimps begin to appreciate how remarkably weak people are. When the chimpanzees made any attempt to interact with researchers, therefore, they were instructed, in Plooij's words, "to act like pillars of salt."

Plooij had spent more than a year watching and assessing the behavior of the adult female Passion and her infant, Prof. But Prof's oldest sister, Pom, could never be avoided, since she was still young enough to spend all day long near her mother. The chimps weren't usually interested in people, and the people had been instructed to act as if they were uninterested in the chimps, so all this scientific observation took place as if an invisible wall separated the watchers from the watched.

One day, though, Pom tried to reach through that invisible wall. Pom approached the young man and began stroking and poking her fingers into his hair. She was trying to groom him, which is a friendly thing to do among chimps. Plooij was astonished but also pleased. He found the sensation of Pom's soft touch in his hair to be "wonderful," and he was strongly tempted to groom her in return. But that would break the rule against interaction. No, to be a proper scientist, to protect himself and the chimps, Plooij knew he must try his best to keep that invisible wall intact, so he acted as if nothing had changed. He remained motionless, unresponding. What else could he do?

Pom, however, continued trying to groom this strange, stubborn ape.

Plooij was now becoming distressed, since it wasn't clear how he could rid himself of Pom. Then he had an inspiration. He remembered how Passion had once discouraged a similar kind of pestering behavior by young Flint. Flint had one time been interested in touching Passion's baby, Prof. Indeed, Flint spent most of an entire day persistently approaching Passion and reaching out to touch the baby; and Passion, who seemed upset and annoyed, could only turn her back to Flint and clutch her baby defensively. In other circumstances, a mother might not put up with this sort of harassment, but Flint was the son of the socially powerful female Flo, so Passion was probably reluctant to smack or chase away Flint because she feared an enraged mama Flo. In any case, near the end of the day, Passion finally came up with a brilliant method for getting rid of pesty Flint. She simply stood up and gazed with dramatic intensity at a distant spot, seemingly watching some especially provocative faraway event.

Flint took the bait. He began gazing into the distance as well, as if trying to figure out what Passion saw that was so interesting. Soon, Flint was moving toward the imagined event, and the second he had moved out of her line of vision, Passion, clutching Prof, just took off in the opposite direction.

So the young researcher Frans Plooij, now remembering this clever little trick by Passion to get rid of Flint, decided he would use the same method to get rid of pestering Pom. He pretended that he had suddenly discovered some astonishing event in the distance. He looked up, gazed intently, even moved his head a little from side to side as if focusing his sight acutely. And it worked! Soon Pom had stopped trying to groom Plooij and was looking in the same direction he was. Pom walked tentatively a short distance toward the imagined point of interest, then looked back at Plooij, who continued gazing. Finally, the young chimp decisively moved off and out of sight, headed into the forest in the direction of that imaginary event, and the researcher was at last able to return, unimpeded, to his observations of Passion and Prof.

A short while later, though, Pom came back, marched directly up to Plooij, and slapped him sharply on the head. Then, for the remainder of the day, she pointedly avoided him.[25]

PART III

What Is Morality? The Attachments

CHAPTER 10

Cooperation

Upon the whole, I thought that the 275th lay would be about the fair thing, but would not have been surprised had I been offered the 200th, considering I was of a broad-shouldered make.

—Herman Melville, *Moby-Dick*[1]

I came to a place of dry grass and a few wildebeests standing in a herd, then I saw a burst of dust, the herd wheeling. A cheetah raged into the middle of the dust, and a single wildebeest was forced away from the rest of the herd. The cat slowed down. The wildebeest drifted away, returned to the herd.

Then I saw a second cheetah and a third, both half-hidden in the grass. They were the same color as the grass only darker, like shadows.

A few minutes passed, and the three cheetahs turned serious, moving away from that ineffectual debate with the wildebeests, moving off in the direction of a small group of Thompson's gazelles. I watched through my binoculars as their movements became purposeful: three cats pacing steadily, heads lowered, shoulders working. One of the three separated and moved off to the right, flanking over to the right of the gazelles—deliberately, I thought, distracting them and encouraging their subtle drift to the left. The other two cats, meanwhile, continued their steady pace, openly parading in a line that cut across the near side of the herd, moving slowly across and, now, separating, with one cat continuing that slow left-flanking movement, the other proceeding slowly but up the center and directly toward the herd.

The gazelles had seemed nervous and sporadically attentive to the wavering grass and the shadows in the grass, but now, as the center cat emerged into full view and boldly approached them straight on, they lifted their heads, and then, just as the cat began to run, they turned and with a flash of white tails bounded off—as the first cheetah on the far right raced in while the other two exploded in full pursuit from the left and up the center.

The three cats converged to pull out a young calf from the rear of the fleeing herd, a baby, really, and now I watched them batting the tiny antelope down. This turned into a distressing vignette of predatory play. It could be they were using the baby as bait to entice the mother. The tiny creature would get up, bawling and wobbly, and try to run away. A cat would trot after him for a few yards, bat him down. He'd cry, *Baaaaa-baaa-baaaaaa.* Then one of the cheetahs carried the baby gazelle in her mouth, put him down—and the baby again struggled onto his feet, cried, tried to run off once more. Now I could see the mother, pulled in by her infant's desperate cries, approaching carefully, tentatively, soon thirty, twenty, then fifteen yards away, placing her delicate self thus in mortal danger. But the mother never came closer, and at last, as she withdrew, all three cats settled down on their haunches and arranged themselves like three polite customers seated around a small table in a roadside diner, as they tore into their shared breakfast. . . .

Spotted hyenas, with their powerful jaws and a digestive system enabling them to consume their prey bones and all, are also masters of the cooperative hunt. These fierce animals acquire the great bulk of their diet through predation, rather than scavenging; and their territorial *clans* can—in areas where prey is especially bountiful, such as Tanzania's Ngorongoro Crater—include up to a hundred individuals.

Spotted hyenas are intelligent, and that intelligence becomes especially evident in the skill and flexibility they show in hunting. During a hunt, different members of the party will take the lead under different circumstances. Individuals will change positions when appropriate, and they adjust their actions and positions according to the actions, positions, and known abilities of their partners. Different prey—with different capacities for defense and different escape styles and tactics—will be hunted in different ways. And the size of a hunting party depends in part on the size and power of the prey. An individual spotted hyena can hunt and kill a wildebeest, but it takes a coordinated team of five or more to bring down a zebra. Many more will join in the predation of a rhinoceros.

For all their exceptionally flexible and coordinated effort during the hunt, spotted hyenas appear violently competitive over the hunt's result, the carcass. Thirty hyenas can consume an entire zebra, including the hooves and skeleton, in about a half hour; and as the size of the feast diminishes, so the feuding over it increases in a way that expresses social rank, with lower-ranking individuals driven away first and the dominant female and her offspring remaining to lick up and chomp down the final bits. This predictable social-class competition over the food means, in turn, that lower-ranking individuals remain choosy about when they will cooperate in a hunt. In essence, they make economic decisions about which hunting parties to join and which prey to chase, since only a large prey with plenty of meat is liable to give them benefits commensurate with the costs of participation.[2]

Animals cooperate in dozens of ordinary activities. They cooperate not only in hunting but in sharing food. They cooperate by forming defensive huddles against bad weather. Some cooperate by taking turns watching out for danger, and by raising the alarm and thereby risking their own well-being when danger does appear. Some cooperate in forming defensive or offensive groups against the threats of predators. Some form alliances within their own social groups: a way of challenging other alliances or individual bullies, upstarts, and competitors. Some share in the care of their young and even share in nursing. Some adopt helpless orphans, and others clean skin parasites from one another, or trade a variety of other services, such as comforting hugs, in exchange for being allowed to touch some mother's baby.[3]

Cooperation among animals is ordinary rather than rare, common instead of anomalous. If we add to cooperation the concept of *affiliation*—friendliness—we find that, moment by moment, act by act, many species seem far more cooperative and affiliative than competitive. If you spend a day with chimpanzees, you may notice some signs of competition, or a reinforcement—made with threats and displays—of an established political hierachy seemingly based on competition. You might even witness an act of violence. But you will more likely see unmistakable signs of cooperation and affiliation. You will probably see mutual grooming. You might be lucky enough to see two individuals holding hands, even, perhaps, kissing in a casually friendly way. You could see one chimp passing food to another. You could see tickle and chase games between youngsters, or between young and old, and you might well observe serene moments of group relaxation, of males and females of all ages sacking out in the noonday sun like sunbathers at the beach. "It is easy to get the impression that chimpanzees are more aggressive than they really are," Jane

Goodall wrote in *The Chimpanzees of Gombe* (1986). "In actuality, peaceful interactions are far more frequent than aggressive ones; mild threatening gestures are more common than vigorous ones; threats per se occur much more often than fights; and serious, wounding fights are very rare compared to brief, relatively mild ones."[4]

Competition will always occur, and it sometimes displays itself openly in violence, which ranges from mild to extreme. As I emphasized in chapter 6, violence is also important, while violence and cooperation are not always opposites. But just as we should not ignore the significance of competition or violence, so we must recognize that cooperation is also a fundamental element in many animals' daily lives.

Scientists and philosophers have often associated cooperation with morality. Some regard cooperation as one of morality's important "strands" or "building blocks": the quintessential "core" of morality.[5] Other theorists have identified cooperation as virtually a full equivalent of morality.[6] Open cooperation, after all, seems to embody the well-known biblical Golden Rule, which, in my modern English version, reads, "So whatever you wish that men would do to you, do so to them" (Matthew 7:12). This is, as we more often think of it, the philosophy of *Do unto others as you would have others do unto you.* I prefer to think of cooperation as one aspect of morality, a sympathetic principle that—along with a few other sympathetic principles or *attachment virtues*, such as kindness and empathy—promotes a group identity that is, in turn, influenced or shaped by moral rules. But whether you regard cooperation as equivalent to morality or merely an important strand or principle, the interesting twist is that cooperation is simultaneously common among animals and yet popularly imagined not to characterize animal behavior at all. Why should there be such a radical disconnect between reality and the popular impression of it?

Two reasons: ignorance and theory.

Our current ignorance about animal behavior is not surprising, since the big revolution in animal watching only began a few decades ago. We still don't know, haven't seen, have not yet documented, an enormous amount. Meanwhile, the recent explosion in what scientists do know about animal behavior has only slowly begun to affect most peoples' general impressions and concepts about the nature of nature. And as long as people remain significantly uninformed about the facts, it is easy enough to subscribe to an unfortunately simplified theory about what Charles Darwin's great idea actually means.

Darwin himself considered that the evolutionary continuity between humans and other animals meant animals ought to have positive social inclinations comparable to those of humans, which under the right conditions would be expressed as morality. He believed the right conditions had to do with high intelligence. As he wrote in *The Descent of Man* (1871), "Any animal whatever, endowed with well-marked social instincts, the parental and filial affections being here included, would inevitably acquire a moral sense or conscience, as soon as its intellectual powers had become as well developed, or nearly as well developed, as man."[7] But Darwin's chief defender and popularizer, Thomas Henry Huxley, articulated an easier-to-digest version of evolution that was to become the powerful majority view. In his lecture *Evolution and Ethics* (1894), Huxley repeated a simple phrase from Darwin's fifth edition of the *Origin of Species* (a phrase that Darwin himself had borrowed from a contemporary, Herbert Spencer), which memorably summarized the concept of evolution as an endless struggle for the Survival of the Fittest. If automobiles had been mass-produced by 1894, SURVIVAL OF THE FITTEST would have made a good bumper sticker.[8]

At the time, people knew precious little about animal behavior, and much of what they thought they knew was wrong. Thus, the idea of evolution as a simple drama of endless competition among individual organisms seemed to explain everything . . . except, of course, for the observed data about a species people did know something about, which was the human one. Among humans, as anyone could see, an evolutionary-style competition could be noted in many consistent and impressive ways, and Huxley's brand of Darwin's idea helped explain that. At the same time, however, Huxley was left with the following dilemma: How could Darwin's great idea ever explain that other side of human existence? I mean, of course, the side where we find people being helpful, kind, caring for one another, following social rules, concerned with social harmony—and, in general, being very cooperative indeed. How could this new theory of evolution explain any of the observed data about the cooperative, seemingly noncompetitive, side of human existence?

The answer, according to Huxley, was that it could not. Evolution explained only the competitive side of the human world. Therefore, so his logic proceeded, one must assume that morality—the cohesive force that supposedly creates and enforces cooperation—is a cultural invention. Nothing to do with evolution. Completely separate from the springs and cogs of our natural machinery. Of course, Huxley's solution to the dilemma was too convenient to be satisfying. It presumes, without ever explaining why or how, that cultural evolution must be far nicer than biological evolution. It leaves us wondering why biological evolution would not, over time, simply grind down the contrary operations of such a cultural imposition. As I've noted in an earlier chapter, it's a deus ex machina solution, desperately cranked down from the stage ceiling, called upon at the last minute because Huxley's carefully cast concept of evolution-as-nothing-but-competition appeared incapable of completing the play on its own.

A reexamination of Huxley's dilemma has taken place gradually, bit by bit, during the last hundred years or so, as our understanding of animal behavior has expanded. The original question remains the same: How can evolution as competition account for cooperation? But the answers have been appearing one by one, in different guises and styles, responding piece by piece to the different forms of animal cooperation scientists have identified.[9]

First came an appreciation of *mutualism*, which might be summarized as *overtly selfish cooperation*. In mutualism, two different individuals interact to give each an immediate benefit. Insects and flowering plants are each primed by evolution to follow their own selfish destinies in a competitive world, yet when an insect pollinates a plant, both receive important benefits. Domestic animals benefit from their relationships with humans by receiving food and shelter, among other things, while humans benefit from their many uses of domestic animals. Humans, come to think of it, have a mutualistic relationship with all those microscopic plant colonies living in their intestines. An Ocellaris clown fish swims freely among the waving tentacles of a Ritteri sea anemone; the anemone's stinging tentacles inhibit clown-fish predators without affecting the clown fish, who has developed an immunity to the stings, while the clown fish chases away anemone-eating butterfly fish.[10] These are a few minor examples of a pervasive cooperative embrace in the living world: mutualism between individuals of different species.

When it happens among individuals of the same species, mutualism may appear in the form of, say, cooperative hunting.[11] The identifying features of mutualism in this case are shared costs, shared benefits, and a minimal passage of time between the two. Three cheetahs come together on the savanna. Each is there for a competitively selfish purpose. But in spite of their competitiveness, the cats hunt cooperatively because chasing Thompson's gazelles by oneself on an open savanna doesn't pay, proportionately, as well as chasing with partners. Better to have a couple of partners who will help trap the prey. The costs? Well, you're required to divide the prey three ways. The benefits might be that you catch more than three times as much prey.

From a theoretical perspective, one attractive thing about seeing cooperative hunting as mutualism is that it doesn't require any great intelligence on the part of the animal. Cheetahs could be stupid—that is, cognitively simple. They don't need to see far into the future or even be aware of much in the present. The promised benefit, food, is right there for all three cats to see and smell and get excited by. The tactics of cooperation could be simple and largely instinctual, the result of a few simple neural algorithms. And each individual gets her selfish reward immediately after the hunt is over.

But what about those spotted hyenas? In this case, the cooperative hunting seems more complicated. The hyenas change leaders, vary positions, alter tactics in response to circumstances and partners. They form different-size groups according to the size and capacities of their quarry. At the end, they divide the carcass in favor of the most socially powerful mothers, who in turn favor their own offspring as well as their siblings and *their* offspring. Some of this cooperation could express mutualism, but if so, the mutualism—joint efforts on an immediate task with quickly shared rewards—is distorted by a hyena's prejudice in favor of relatives.

We humans are familiar with mothers giving their offspring the best place at the table and with relatives quietly favoring each other in the workplace, and we call this *nepotism*. Nepotism is ubiquitous in traditional human societies where power is inherited, where princes become kings and princesses turn into queens, but we also expect it even in nontraditional societies where power is supposed to be distributed according to merit. Why else, when he had hundreds or even thousands of perfectly qualified candidates to choose from, did U.S. president John F. Kennedy select his brother Bobby for the post of attorney general?

Nepotism is so much a part of the human experience that we seldom question its logic, even on those occasions when we dispute its fairness. *Of course!* we say. *Isn't it obvious? He's family.* But it is an asymmetrical form of cooperation compared to mutualism, since in mutualism both participants benefit, whereas in nepotism one individual incurs a cost so that another may benefit. The cost may only be a lost opportunity to select the most competent person for an important job because it was given to a less competent relative, or it may be a clear and undeniable cost, as when a mother risks her life for a child or an uncle goes to jail because he's protecting a nephew.

That nepotism is so common among humans as to have become invisible—obvious, just the way things are—suggests that it may be part of human nature, a significant inclination placed in the human psyche by the workings of evolution. And when we shift our attention from humans to animals, we find much more of the same. The cooperation based on nepotism that we observed among spotted hyenas is not unusual. We can discover nepotistic cooperation in a large number of animal societies. If it still seems obvious and natural at first, the real puzzle emerges once we discover those many instances where the costs to one and the benefits to others are starkly contrasting, as with Belding's ground squirrels.

These burrow-dwelling, earthbound squirrels live in the high, open meadows of far western North America. They hibernate during the colder months, but when the sun and thaws arrive in late spring, they emerge and begin their active lives. In the Tioga Pass Meadow of the Sierra Nevada mountains of northern California, where they've been thoroughly marked (ear-tagged, toe-clipped, and hair-dyed) and extensively studied, the summering squirrels suffer from regular attacks by four wild species—long-tailed weasels, badgers, coyotes, and pine martens—as well as the occasional domestic dog unattached to a human master. They are also under threat from winged predators overhead. The squirrels' primary defense against predators is to hide or escape down into their burrows, and they often benefit as a group when one individual—not necessarily the first to see the predator or even the one nearest to an approaching intruder—gives the alarm call.

A predator flying overhead is identified with a single-note whistle. For four-legged predators approaching on the ground, the alarm call is a staccato burst of short, high-pitched chirps or cries. The call is repeated, and those repeats will continue for a few minutes after the intruder has disappeared from sight.

The squirrels who make a call take a significant risk. The caller will typically sit bolt upright on a high rock and turn to face the approaching beast. The head-on position of the caller may alert the others not only to the presence of a predator but to the predator's position; soon the others will have oriented themselves in the same direction as the caller, as they prepare to flee in the most appropriate direction. So alarm calling is a real service to the community, but the caller herself is two to three times more likely than the other squirrels to be stalked or chased by the predator. That translates directly into a higher mortality rate. During a three-thousand-hour study conducted at Tioga Pass Meadow in the early 1970s, zoologist Paul Sherman and his assistants observed terrestrial predators kill nine ground squirrels, six of them adults. Of the six adults, half had been the same ones who gave the alarm.

Here, then, is a case of cooperation in which an individual provides genuine benefits to others while doing so at a very real cost to herself—risking her own life. Why? Why, if we adhere to Thomas Henry Huxley's concept of evolution as Survival of the Fittest, should evolution ever produce—not accidentally, but regularly, routinely, as a matter of course in a stable system—a species where some individuals risk their own lives so that others might live? Who, in a world ruled by nothing but selfish competition, would ever unselfishly choose to be first to bring dangerous attention to herself by giving the call?

I have been speaking of the alarm caller as *her* for a reason: Adult females ordinarily give the calls. Adult males seldom do. This is true in mixed-sex groups, and it plays out in single-sex groups as well. When predators approached all-female groups during Sherman's study, some individual gave an alarm call in 85 percent of the instances, whereas when predators approached all-male groups, one of the males gave an alarm only 18 percent of the time.

This sexual disparity in alarm calling is echoed by a sexual disparity in kinship associations. As evolution's way of discouraging incest and inbreeding, females stay in the meadow of their birth while males leave before their first hibernation to find a more interesting meadow beyond the horizon. Once the females reach maturity, they become sexually receptive for around four to six hours during the breeding season, and the males who then mate with them are strangers who came from somewhere else. Only a few males will mate with most of the females in an area, and by the time the females have given birth, those sexually active males will also have left the area. The females nurse and raise their young alone, without help from any of the males. In short, adult males are seldom near their genetic relatives, while the females are typically surrounded by them—by their offspring, their mothers, sisters, and aunts. Why should females give the alarm call much more often than males? Probably, Paul Sherman concluded in 1977, because they are watching out for their relatives.[12] That's cooperation through nepotism.

Evolutionary biologists had been thinking about the principle of nepotism at least since the 1930s, when J. B. S. Haldane cleverly declared, "I would lay down my life for two brothers or eight cousins."[13]

Haldane was particularizing the genetic relationship between individuals. If we examine that part of the genetic code that normally varies fully (or 100 percent) between individuals in a species' gene pool, we find relatives to have markedly less variance than average. Siblings are around 50 percent genetically nonvariant, while cousins only share an average of 12.5 percent of their variant genes.[14] In thinking about relationships this way, as Haldane did, we are recognizing an important principle that marks a second solution to Huxley's dilemma. Yes, evolution is always and invariably marked by competition—the old battle known as Survival of the Fittest—but it is competition, survival, and fitness at the genetic rather than the individual level.

Looking at the world from a gene's perspective, we can imagine individual organisms as nothing more than gene carriers. At this level of thinking, then, the chicken is servant to the egg. Once we imagine the individual organism as a gene carrier, we can recognize the gene's task in its starkest form. Gene carriers die. Genes don't have to. The gene's task is to survive over time by placing itself in the care of the next generation of gene carriers. If the gene does that, it continues to exist, even thrive. If it fails, it will go extinct. Obviously, genes don't experience feelings or thoughts or anything else that would allow them to recognize what their task is or is not. They don't need to. Any genetically based quality that makes the gene carrier reproduce more successfully than average will expand its gene's representation in the next generation. Any genetically based quality that makes the gene carrier reproduce less successfully than average will help push the gene toward oblivion. In that way, genes are increasingly shaped by evolution to define qualities that promote the successful reproduction of their carriers.

This might seem to you just a byzantine way of rephrasing Huxley's Survival of the Fittest slogan, but recognizing evolution as competition among genes rather than among individual organisms can clarify our thinking, making it both simpler and more expressively complex.

In seeing evolution as a competition occurring among genes rather than among individual organisms, we can revise our expectation of how evolution might work in individuals. We can imagine, for example, that the selfishness that marks evolutionary competition could characterize the gene yet not the gene carrier. Selfish gene may not equal selfish person—or, for that matter, selfish elephant or selfish dog or selfish Belding's ground squirrel. It might be possible for the competition among genes to advance through genuinely cooperative individuals. Certainly, at least, for individuals who are genuinely cooperative with their relatives.

Mutualism and nepotism explain a lot, but we are still left wondering about other, seemingly more complex instances of animal cooperation, especially those involving nonrelatives. Consider, for example, the sophisticated hunting styles of West African chimpanzees living in Ivory Coast's great Taï Forest.

Among these chimpanzees, young males can join in the hunt only after they've reached the age of eight to ten years, but the skills shown by the best hunters are acquired over twenty years. The Taï chimps prey mainly on red colobus monkeys, who have the advantage of being four to five times lighter than the apes. This means that the monkeys can move onto fragile branches and other parts of the high forest canopy where apes cannot. The chimps compensate for such a disadvantage with their great strength and sharp intelligence, and by organizing their hunts.

Upon spotting a promising troop of red colobus monkeys in the trees, a hunting party of chimpanzees approaches stealthily from the ground until they reach an area beneath the still unsuspecting prey, at which point one of them begins to climb, slowly, up to a height of about five meters. The climbing ape is not intent on capturing a monkey. Rather, he works as a *driver*, gets the monkeys spooked and moving in a particular direction. Even before the monkeys see the driver and begin to flee, though, the rest of the hunting party on the ground has begun moving in anticipation of where the monkeys will go. When the monkeys are alarmed and moving away from the driver—climbing higher, rushing along various branch highways, leaping between branches, and diving from one tree to the next—the other chimpanzees work to focus that movement. Some of the chimps serve as *blockers*, doing nothing but anticipating a possible escape route and blocking that route by climbing a tree and remaining in that single position. Some join in as drivers, climbing up and keeping the troop moving in the right direction. Others now work as *chasers*, isolating smaller groups of individuals or perhaps a slower-moving mother burdened by her clinging infant. The chasers might actually now catch a monkey, but it's unlikely. The fleeing monkeys are agile and light enough to stay in the higher tree canopy, while gravity forces the much larger chimps to remain in the lower and more reliable parts of this complex three-dimensional maze.

Still, the drivers, blockers, and chasers have so far managed to get the colobus monkeys in motion, have isolated a more vulnerable group from the larger troop, and have begun to direct their flight. Up ahead, anticipating not only the future direction of the isolated monkeys but their probable height off the ground as they flee, a final chimpanzee—the *ambusher*—has posted himself at a certain height in a certain tree. The ambusher is most probably an older and experienced hunter, since his critical role requires a well-developed concept of how monkeys do things in order for his double anticipation (exactly where and how high) to succeed. When the monkeys do arrive, the ambusher leaps out from an enveloping cloud of leaves and forces them to turn back and, desperately now, to descend into the lower and sturdier parts of the forest where the apes have suddenly gained an advantage in speed and maneuverability over their prey—and so an unlucky monkey is, at last, agonizingly transformed into food for apes.

The Taï chimpanzees cooperate in their hunt with nonrelatives, and they take on various positions and roles with no obvious signs that there will be a commensurate reward waiting for them at the end. Meanwhile, the same males known for cooperative hunting are also known for sharing their meat with females and males too young to hunt. How much meat? Taï females rarely hunt, yet they eat an average of twenty-five grams of meat each day. These chimpanzees also maintain the cultural tradition of using stone and wood hammers and anvils to crack open hard forest nuts. Nuts are a second important source of protein, after meat, but cracking them open still requires a polished enough technique that the young need help, since they only slowly acquire the necessary skills through imitation and active learning. Thus, adult females, while providing for their own daily intake, also share a significant amount of unshelled nuts with the young of both sexes.[15]

That chimpanzees share food is interesting, you might think—but then, you may say, chimps are so darned clever they break most of the animal rules anyhow. Right? Wrong. The truth is that many species share or exchange goods and services, and in doing so they may be cooperating in a third style, known as *reciprocity.* Chimpanzee food-sharing is partly based on reciprocity . . . but perhaps the blood regurgitations of vampire bats make a clearer example.

Common vampire bats have short ears and small, cone-shaped muzzles with specialized heat sensors at the tip, enabling this creature to discover areas on the skin of prey where blood flows closest to the surface. Like other bats, they hunt at night and navigate in complete darkness through echolocation, or sonar. This particular species, though, having evolved to feed on the blood of mammals, has developed a special sensitivity to the distinctive sounds produced by the slow, in-and-out breathing of a deeply sleeping domestic cow or goat or other mammal, including, on rare occasions, a person. Having located a good source of blood, the bat lands, approaches cautiously by scurrying on the ground, and finds a likely spot on the prey's body for reaching the blood. If hair or fur is in the way, the bat uses her canines to shave an area clear down to the skin, and then, with her sharp upper incisors, she penetrates the skin while injecting an anticoagulant saliva.

A female vampire bat weighing around forty grams can drink half her body weight of blood in about twenty minutes. It's good to drink fast, in case the source of that blood should wake up in a bad mood. However, all that extra weight could make it hard to fly away. Thus, evolution has given this winged mammal a digestive system that quickly transforms blood's nonnutritious liquids into urine, and she will start to expel the urine within a couple of minutes after the drinking has begun. By the time she's ready to fly away, the bat has actually increased her body weight only by a fifth to a third—still light enough that, with an extra crouch and a flinging leap, she casts herself into the air and swoops off to the roost, where she will spend the rest of the night with the colony: around a dozen other adult females plus their young of both sexes and one adult male.

Blood ought to be easy to find, but vampire bats under two years old fail to feed in one night out of three, while adults come back hungry in about one out of ten. Hunger represents not merely a gnawing discomfort, but the potential for a fast downward spiral proceeding into weakness and a reduced ability to forage for blood, and ending, after three hungry nights, in death by starvation. It might make sense, then, to share food on occasion, which the bats do with mouth-to-mouth regurgitation, a transfer of partly digested blood from the full to the hungry that superficially resembles kissing. One bat returns to the roost hungry after a night of futile blood-hunting. Another sidles over and provides the kiss of life.

University of Maryland zoologist Gerald S. Wilkinson, who has studied vampire bats in Costa Rica, reports that during a hundred hours of observation inside several roosts, he saw food sharing through regurgitation on

110 occasions. Seventy-seven of those regurgitations were done by mothers feeding their dependent young, while thirty-three happened between two adults or adults and juveniles. Considering just those thirty-three regurgitations not involving mothers and their young, Wilkinson found that the partners were sometimes close relatives and sometimes unrelated but preferred roostmates.

Wilkinson also noted that the adult females spent around 5 percent of their roosting time grooming each other in a pattern that focused on the same close relatives and preferred roostmates. Unlike a lot of mammalian mutual grooming, moreover, this did not seem to be an exchange of services—removal of one another's skin parasites, for example—but rather a series of mutual examinations that focused on the partner's stomach: *How's the tummy today, my dear?* A vampire bat's stomach becomes notably distended after a good feed, so stomach examination could be a way of determining a bat's recent feeding history, determining who's hungry and who's full. It may also be, so Wilkinson believes, a way of keeping score and identifying cheaters: those bad bats who have received blood regurgitations when they were in dire need but who now, when they're rotund and obviously full, refuse to share.[16]

Wilkinson believes those blood-trading vampire bats provide a good example of reciprocity partly because in a significant number of cases the blood regurgitations occur between unrelated roostmates. We can understand why relatives might share blood when one is full after a satisfying night of blood-sucking while another is weak from hunger, but why nonrelatives? The answer is reciprocity. The unrelated roostmates trust one another to return the favor.

Reciprocity among animals resembles economic activity among people. The underlying principle is the same: *You scratch my back, and I'll scratch yours. Give me something, and I'll return the favor.* It differs from nepotism in that the partners involved aren't necessarily genetic relatives. It differs from mutualism largely because of the time factor. The costs and benefits are not given or received simultaneously, which means that the participants have to keep track of what's going on. They have to keep score. One individual begins the exchange by incurring a cost for the benefit of a second individual. Time passes. Then, ideally, the second individual returns the favor, either in similar or dissimilar form. I say "ideally" because, of course, the second individual might default or cheat. Uncontrolled cheating could make cooperating less desirable than not cooperating. Reciprocity is how cooperation can work in a universe of unrelated egoists, but it requires that we discourage cheating by keeping score and punishing the cheaters.[17]

Reciprocity—with its scorekeeping and punishment of cheaters—may still seem exotic to you, and probably the example of blood-vomiting vampire bats makes it look even more exotic. The principle is not hard to understand, however, and actual instances of reciprocity are widespread. When you watch chimpanzees at a zoo or go see them in the wild, you will sooner or later find a couple of chimps sitting side by side, or one behind the other—or maybe three or four or five of them similarly arranged close together in various strange postures and attitudes—and picking through each other's hair. Hair on their arms, backs, stomachs, heads, necks, and so on. This is reciprocal or mutual grooming, and it looks a lot like the kind of grooming we humans pay for when we go to the barbershop or beauty parlor. In our case, the reciprocity is not an exchange of grooming for grooming, but rather one of money for grooming. We exchange money, our symbolic promise of future benefit, for the immediate and practical benefit of having our hair cut and, ideally, our appearance improved—and also, often, for the impractical benefits of gossiping, chatting, sympathetic agreements on baseball and politics, and a general warm feeling that comes from simple physical and social closeness with another person. With chimps, grooming includes the important practical benefits of removing skin parasites, and the impractical (or practical but less obviously so) ones of being close to another chimp, reaffirming a friendship, making up after a conflict, strengthening an important political alliance, and so on.

Mutual grooming is a nearly literal example of reciprocity as you-scratch-my-back-and-I'll-scratch-yours, and the exchange happens only because, as in any other economic activity, the whole is greater than the sum of its parts. That is, both participants are receiving benefits greater than what they would expect to gain with the same effort on their own. It is possible, of course, to stroke yourself and remove fleas, ticks, and other parasites from your own arms and legs, but only an attentive other chimp can find and remove the parasites lodged in the back of your neck or on the top of your head. And parasites are not merely an irritant, but a genuine health hazard. The higher your level of parasites, the lower your fitness—that is, your likelihood of surviving and reproducing.[18]

Primates in general—apes, monkeys, and prosimians—are well-known for their mutual grooming, but impalas may provide the neatest example of it. Among impalas living in East Africa, a typical grooming session begins when one animal turns to another and delivers a single bout of six to twelve parasite-removing tongue-licks and tooth-scrapes to the chosen partner's head or neck. The other impala then turns to the first and delivers an approximately equivalent bout of licks and scrapes. This goes on, back and forth, bout for bout, tit for tat, for an average of six to twelve bouts per session, and then the session is over.

The reciprocity of mutual grooming among impalas is remarkably simple. One animal starts it. The other responds. So it goes like the tick-tocking pendulum on a grandfather clock. This is true no matter whether the grooming occurs between males, between females, between males and females, or between old and young. The males living in bachelor groups are unrelated, while females living in their own groups are only mildly related—but genetic relationships don't seem to matter. Males in all-male bachelor groups develop a dominance hierachy that females do not have, and a grooming session between males will usually end with a little pressing of horns together or a light sparring with horns. But the sparring never leads to serious battles, and the dominance pattern is not reflected in the pattern of grooming.[19]

For impalas, the scorekeeping must be pretty easy. It could be automatic. Still, if scorekeeping is not done at all, the system would likely collapse—that is, become evolutionarily unstable—because of cheaters. Scorekeeping is how we uncover cheaters, and cheaters must be discouraged for reciprocity to continue as an evolutionarily stable process.

Cheating impalas? Consider, for instance, a rather lazy fellow who makes a habit of allowing others to initiate grooming and is always the first to stop. This habit means that in every partnership, if one impala provides nine lick-and-scrape bouts, our lazy impala only returns with a total of eight. Our lazy impala is therefore shortchanging the others. It may be trivial enough that no one will ever notice; if so, our lazy impala is getting the benefit from other impalas' normal nine-bout grooming sessions, while he spends a little more than average time eating, watching out for predators, and keeping an eye open for good opportunities to mate. If we assume that those minor benefits of being lazy are even slightly significant on average, in a lifelong evolutionary competition with others, then we can predict that over time, perhaps over many hundred impala generations, natural selection will produce a new set of impalas who are on average all a little lazier, all licking and scraping each other an average of eight bouts per session rather than the previous nine. But then, of course, the same thing will continue to happen. Because of cheaters, or consistently lazier-than-average impalas, the average number of grooming bouts per session could decline until mutual grooming disappears.

Reciprocal cooperation benefits both parties, but cheating—in which one suddenly uncooperating party takes an even greater benefit than a cooperator gets—can undermine it. I believe it is impossible to eliminate cheating, just as in a human economy we can't expect to be completely free from the depredations of thieves or embezzlers. But it's not necessary to eliminate them. One only has to discourage or punish them enough, just enough, that honest cooperation is at least slightly more beneficial, on average, than cheating. We watch out for cheaters by scorekeeping. Punishing cheaters could range from obvious acts—a kick in the ribs, say—to a dozen or a thousand more subtle forms of punishment including mild social isolation.

I say these things based on the theory of reciprocity famously presented by evolutionary biologist Robert L. Trivers in 1971.[20] In fact, the study of impala grooming I've been referring to here only documents their clear reciprocity without considering whether or how the impalas are keeping score, whether or not there are impala cheaters, and whether the cheaters are somehow kept in line through some sort of punishment. So I am just theorizing that these impala ought to be keeping score, and I am only speculating that they somehow punish cheaters. Keeping score, though, can be more simple and direct than it might seem in theory, while the punishment of cheaters could be subtle to the point of invisibility. Take the case of lagging lions.

Female lions in a pride form cooperative groups to defend their territories against the aggressive incursions of outsider males and outsider females. The defensive ability of these cooperative groups can have life-or-death consequences, yet some of the female lions lag significantly behind the front lines, thus subjecting the others ahead to greater risk than necessary while keeping themselves relatively safe.

A pride of lions typically consists of three to six individual adult females, but it can include as many as a dozen or a dozen and a half. In any case, a pride includes a small enough number of adult females that they can all know and recognize each other individually. The laggards are typically the same ones every time. Their lagging appears unrelated to size or fighting ability, and since the females don't form obvious dominance hierarchies, it seems unrelated to rank. Laggards are just laggards, and their behavior shows up at an early age and remains consistent over a lifetime. Laggards are obvious, and you can almost see the scorekeeping that takes place. Every time a pride goes out in a group, assembling to present a united front against a real threat, such as an intruding pride of stranger females, the laggard is reluctantly walking way behind the others. Meanwhile, the leaders in this group often pause and glance back at the laggard, as if to say, *Come on! We need you.*

That's scorekeeping, and it seems simple enough—but are the laggards really cheaters? Trivers's 1971 theory of reciprocity imagined a beautifully binary world of cheaters versus cooperators, whereas little in reality is so simple. One might say that the lagging lions are just imperfect cooperators. Or maybe they're just partial cheaters. Or maybe we should say they're sort of cheaters and sort of cooperators. After all, even though they're lagging, they are still also available to defend the rear flanks of the advancing pride, or to move forward at a critical moment, so their position near the rear is not useless. And the fact that scientific observers have yet to see a lead lion stride back to the rear and mightily cuff one of the laggards does not at all demonstrate that the laggards go unpunished.[21]

Think back, for a second, to your own experience with laggards. Maybe you were one. Maybe, in your childhood games, you simply could not muster the full enthusiasm the others seemed to have as they chased that ball around for a couple of hours, meanwhile yelling and tripping over one another and getting profoundly sweaty. It was fun, on occasion, but for you not as much fun as the others seemed to be having. So you played with an enthusiasm that didn't reach the average level. You were, in short, a laggard. No one accused you of being one, but then no one had to. You knew you were an imperfect cooperator, as did everyone else. The punishment came on the next day, when everyone got together once more to choose up new teams, and you were chosen last of all. How humiliating!

Your punishment could have been almost completely invisible to an outside observer. No one cuffed you or even scolded you. You were just made to feel less welcome or less important in the group. Your punishment was an emotional one, a smack upside the head from the inside out: one, that is, meted out by the neurochemistry of your own brain—and as subtle as it might have been, you can still to this day remember the pain.

Meanwhile, you also missed out on the neurochemical pleasures happening inside the brains and bodies of the leaders of the group, the ones who were absolutely caught up in the passion of the game and so experienced some positive feelings of inclusion and power and are now feeling quietly proud because they were chosen first. Your punishment, then, is both to feel bad for being excluded and to miss feeling good in the way that the fully cooperating players feel. This means that nature, working through evolution by natural selection, has created a neurochemically based series of punishments and rewards to keep you and your friends playing the game.[22] The punishments are not enough to make you leave the game altogether; and for you, given your particular physique and temperament, the potential rewards are still not enough to turn you into a fully committed cooperator. What they do instead is make you try a little harder or at least discourage you from becoming even more of a laggard. The net result is that the balance between leaders and laggards in this game is stable enough that the game goes on, much as it has always done in a past beyond remembering.

Why should nature care? Why should evolution have introduced such a complicated neurochemical system of rewards and punishments to buttress reciprocal cooperation in what is, after all, a rather silly children's game? Why? Because children grow up. Because that silly children's game can become, with a few minor alterations, a serious adults' hunting party or an advancing coalition of male or female warriors ready to defend the full community against an invading coalition of outsiders.

Herman Melville's potrayal of life aboard a nineteenth-century American whaling ship imagines an all-male human society where the participants, none of them genetic relatives, cooperate during nearly all their waking hours. Each of the sailors has placed himself aboard the ship, laboring day by day, spending energy and risking life, contributing to the voyage and its economic potential, because each whaler is being paid for his efforts, and the payment is actually a promise of some abstract bit of wealth in the future. Each sailor's reward is to be determined according to predetermined portions of the ship's final profits, and the portions are called *lays.* It is impressive how precisely the men involved try to calculate what their lays should be.

Ishmael, the book's main speaker and central character, signs his own shipping papers after an extended bargaining with the owners of the ship over which lay he fairly deserves. This preemployment negotiation starts in his own mind with an exaggerated sense of his own fair worth. He begins by considering that, as a common seaman with no experience, he should be worth about one 275th of the profits—a 275th lay "would be about the fair thing." But then, taking into account his sturdy build, he quickly readjusts that calculation to something more favorable: one two-hundredths of the ship's profits, or the 200th lay. Ordinary egotism makes a person exaggerate his or her own fair value, and Ishmael's naive calculation is comically dashed by the pair of seasoned and contentious shipowners, one of them a grizzled Quaker named Bildad. Bildad insists that it would be fair indeed to give this green sailor merely a 700th lay. After a good deal of disputatious consideration, though, the owners arrive at their final offer of a 450th lay, which Ishmael at last agrees is fair enough. He signs the papers formalizing the arrangement, and then, officially a member of the crew, he prepares to board the ship.

If we look at the ship and its crew of a couple of dozen men as a simple problem in reciprocal cooperation, we can see what an astonishingly complex thing it is. Two to three dozen men, each with his own distinctive needs, talents, temperament, education, and abilities, join together and merge with the ship, cooperating almost as if they were all part of a single superpredator that goes onto the raging seas to kill an animal who is vastly more powerful than all the men put together. Put together, that is, without the multiplying effects of cooperation. If we recall that each man has agreed to cooperate based on the promise of a certain lay to be paid out in the blissfully hazy future, and each man is expected to cooperate with all of the other men in the painfully clear present, then we might conclude that the sheer intelligence required to do the math, to remember everything, to keep score and watch out for the cheaters, is just remarkable indeed. But, of course, things don't work that way. It doesn't really take that much of any single person's intelligence to run this ship, because the cooperation and scorekeeping are, in large part, distributed and done automatically with the help of language and money.

Language and money are the symbolic systems that have helped humans become the most cooperative species on this planet. They are also the systems that allow Ishmael and the ship's owners to sit down before the voyage and negotiate over a fair lay to be paid after it's over. That is an economic exchange, the buying and selling of labor, the trading of one benefit for another. With spoken language, the terms of fair reciprocity are first negotiated and agreed upon, and then with written language, using quill and ink and paper, they are given some measure of stability over time. At the successful conclusion of the voyage, so Ishmael fervently hopes, those ink-on-paper scratchings will be translated directly into money, which can then be used in further reciprocal exchanges with other people for more goods, services, and adventures.

Animals don't have language and money, but they reciprocate and keep score with other kinds of radical shortcuts. Indeed, keeping score itself happens only because the process is not so much a complicated intellectual one—not a matter of counting and remembering the numbers—as it is a quick emotional one, an innate feeling of rightness combined with a learned sensation about what is the appropriate degree of effort in some regular exchange. We can call that learned sensation and the unlearned, automatic feeling of rightness our *sense of fairness*. Ishmael himself describes his estimates about which lay he should receive as a search for "the fair thing."

A sense of fairness does not require formal counting, arithmetic, or any other act or experience involving complex cognition or an appreciation of others as independent agents.[23] The sense of fairness merely requires some feeling about the passage of time or the passage of an expected sequence of behaviors in a particular reciprocal association. *It's my turn now.* That's what both my dogs, Smoke and Spike, tell me once in the morning and once at night, when they start squeaking and whining and pacing, shaking their noisy collars and giving me their significant gaze. Their timing is almost good enough to set the clock by, and their emotional expressions of expectation are unmistakable. *It's our turn now for a walk*, so they would say if they could speak the words. It's their turn in our reciprocal relationship—a reciprocity they don't talk about, haven't really thought about, don't understand intellectually, but are psychologically attuned to nevertheless. They will have their walks, twice a day, and in return, in reciprocation, they will do the many things that I expect from them. It's only fair.

It's my turn now. One can now imagine how the scorekeeping of mutually grooming impalas operates. An impala doesn't keep score by counting the other impala's licks and scrapes. Impala reciprocity is not evidence of an exceptional brain or the capacity for understanding numbers and doing arithmetic. Rather, it's a sign of the simple ability to sense a certain rhythm, a particular passage of time, and the arrival of a particular degree of tongue and mouth fatigue. At the end of it all, there comes the feeling that this effort has been enough, and so the sensation of fairness arrives. "It's my turn now," our hypothetical impala says, "for my own licks and scrapes."

So an intuitive sense of fairness is a way to keep score for animals, as for humans.[24] A second way of keeping score is to be found in special relationships.[25] In one free-ranging troop of baboons living in the Okavango Delta of Botswana—including nineteen adult females, some relatives, some not—a recent study found that the average female devoted about 95 percent of her grooming time to only four other females, while nine out of the nineteen females actually spent more than half their time concentrating on a single grooming partner.[26] The common development of special relationships—alliances and friendships—among these baboons makes scorekeeping vastly simpler than it would otherwise need to be if we imagined the alternative scenario of each female trying to exchange grooming with all eighteen of her fellow females in the troop.

Alliances and friendships define relationships where the scorekeeping is simply less critical than it would otherwise be, since alliances and friendships have a past history of satisfactory reciprocity. This satisfactory past brings into the present a relaxation of the need for scorekeeping because, so we say, *trust* has been established. Trust is a reasonable response to past experience, but it may be facilitated by the presence of a compound called *oxytocin*, a neuropetide with physiological functions in childbirth and lactation but also associated with pair-bonding and maternal care, as well as sexual and social attachments. Oxytocin in animals enables them to overcome their normal inhibitions about physical proximity, and it thus encourages the development of trust-based alliances and friendships.[27]

Ishmael's best friend aboard the *Pequod* is an elaborately tattooed South Pacific islander named Queequeg, a brave and powerful man who is possibly a cannibal but certainly a first-rate harpooneer. Much of the cooperative activity Ishmael engages in aboard the *Pequod* is actually based on his trusting friendship with Queequeg. They are, as the other sailors like to say, "chummies." They work well together, just as all the other sailors on board work best and most often with their appropriate allies and particular friends. The cooperation that enables this ship to sail and the whalers to hunt whales, then, does not resemble a well-woven blanket: two dozen men cooperating as one superbeing, all of them mutually exchanging their pushes and pulls, heaves and hos. Rather, it is a complex network of men paired up in limited, trusting partnerships that exchange with one another their pushes and pulls, heaves and hos. A network, not a blanket. So, too, one can predict, will the reciprocal cooperation appearing in animal societies of a certain size break down into a network of individuals reciprocating with individuals to provide cooperative results that eventually, in the spreading style of any network, expand broadly to benefit the larger community.

CHAPTER 11

Kindness

The monkey-rope was fast at both ends; fast to Queequeg's broad canvas belt, and fast to my narrow leather one. So that for better or for worse, we two, for the time, were wedded; and should poor Queequeg sink to rise no more, then both usage and honor demanded, that instead of cutting the cord, it should drag me down in his wake. So, then, an elongated Siamese ligature united us. Queequeg was my own inseparable twin brother; nor could I any way get rid of the dangerous liabilities which the hempen bond entailed.

—Herman Melville, *Moby-Dick*[1]

One day in late June 2000, a young African forest elephant, weak from malnutrition, collapsed off to one side of a narrow, sandy trail in a Central African forest. She lost consciousness and within a few hours died, and her collapse and death were witnessed by her mother and older sister.

Since this trail was an important route from the forest out to a swampy clearing with minerally nutritious muck, reliable drinking pools, and attractive mud wallows, it was well traveled by elephants. The trail was also close enough to a high observation platform erected at the edge of the swampy clearing that scientists sitting on the platform could watch the dying elephant through their binoculars. Using a video camera with a telescopic lens, they also documented the reactions of a large number of elephants as they ambled along the trail to and from the swampy clearing.

During 5.25 hours when the elephant was dying on the first day, and during another 6.5 hours after she had died on the second day, the video camera recorded a large number of elephants passing by, and using that record, the scientists analyzed the reactions of elephants to a fellow elephant in trouble. Since all of these elephants had been studied for years by Andrea Turkalo, an expert on forest elephants, the scientists also had access to identification records, which enabled them to consider the sorts of social and kinship relationships, if any, that the passing elephants had with the dead elephant's mother, Morna, and sister.

During the first day, thirty-eight elephants walking along the trail made a total of fifty-six visits to the elephant who lay dying. Six of those visits were made by Morna and the older sister. By the end of the first day, after the young elephant had died, the mother and sister left the area, but another fifty-four individuals made seventy-three visits on the second day. During those two days, then, elephants walking along the sandy trail made 129 visits to a fellow elephant in trouble. Of those 129 visits, a total of 128 showed some kind of distinctive response to the dying and then dead animal.

Most of the responsive visitors began by exploring the body and the area around it, curiously sniffing the air, gently touching the body with their trunks, probing the body with their feet, and so on. But after that inital exploration, how did they react? About 50 percent of them reacted as you might expect: They showed signs of fear or avoidance. An elephant might back or sidle away from the body, for example, or move off the path or run off hastily after a brief exploration. Although this particular spot (Dzanga-Sangha *baï* of the Central African Republic) is within a legally protected forest, poachers had regularly been hunting elephants for meat and ivory throughout the larger region, so for any elephants walking along the trail, a dying or dead elephant alongside the trail could mean serious danger.

One exceptional individual, known as Miss Lonelyheart, visited several times on the second day and reacted aggressively to the body, stabbing it with her tusks and attempting to tear pieces away from it. Miss Lonelyheart was already well-known as a social misfit, and her bizarre behavior was not out of character. But that 50 percent of the visitors who reacted with fear and avoidance: Who can blame them? For an elephant even to take the time to stop and investigate another dead elephant amounts to expending extra energy and taking an unnecessary risk. How remarkable, then, was the behavior of the elephants identified in the other half of the sample, which included many instances of socially positive reactions to the drama of another elephant in trouble. Some 15 percent of the total visits during those two days involved protective behavior: the visitor seeming to protect or guard the body from others. And in about 18 percent of the cases, the visiting elephants looked as if they were trying to assist or revive the dying elephant, mostly by attempting to push or lift her upright, using their feet, trunks, or tusks.[2]

The researchers found no correlation between protective or assisting behavior and genetic or social relationship. The visitors were not at all limited to friends or associates or relatives of Morna and her daughter. Even when the scientists ranked the visitors according to the intensity (measured in time, number of visits, and number of helping acts) of their responses, they still found no pattern that suggested either kinship or previous social relationship was important. These apparent acts of kindness, then, might be examples of true altruism.

One famous story about altruism among humans bears an interesting resemblance to this account of elephants. I'm referring to the parable attributed to Jesus and described in the book of Luke, chapter 10: the story of the Good Samaritan. It goes like this.

A man traveling on the road between Jerusalem and Jericho was set upon by robbers, who took everything he had, stripped him naked, beat him severely, and left him to die alongside the road. As the man lay there, dirty, naked, bleeding, helpless, dying, three people passed by and noticed him.

The first two were seemingly upright and respectable men, well-established members of the same community of Jews being addressed by Jesus as he tells the story. When the first man, a priest, saw the unfortunate victim lying there, he crossed to the other side of the road and continued on his way. The second man, a Levite, also crossed to the far side of the road and moved along. These were apparently honorable men, and their behavior still seems reasonable. Perhaps they were worried about the presence of thieves. Maybe they were in a hurry for various good reasons or were appropriately reluctant to dirty themselves or to take on the several obligations that would inevitably appear once they began attending to the needy man.

The third person to come along was a Samaritan, a member of that despised and distrusted sect of apostate Jews who had developed their own version of the Torah. What would a Samaritan know about goodness or kindness or personal ethics? When this man saw the robbers' unfortunate victim, however, he stopped and attended to him. He washed the man's wounds with wine and soothing oils, wrapped them with bandages, and then placed the man onto the back of his own pack animal and transported him to a nearby inn. The Good Samaritan paid for the victim's lodging there, and, upon leaving the next day, he placed two silver coins in the hands of the innkeeper to pay for the victim's care and promised to pay for any additional charges when he returned.

This didactic tale provokes us to consider our duty to our neighbor. Our duty is to behave not like the two respectable men who were absorbed in their own concerns or inhibited by their fears, but rather to emulate the third one, the Good Samaritan, who dramatized one of the central ethical principles of Christian teaching, which is to practice radical kindness: to "love . . . your neighbor as yourself" (Luke 10:27).

I think of *kindness* as the spontaneous response to someone else's need or suffering. If that someone else is a close relative, then such kindness is neither unusual nor surprising. A baby cries, and we can be sure the baby's mother will quickly respond, doing all she can to resolve whatever distress made the baby cry. No surprise there—either in terms of common sense or evolutionary thinking. The baby of the mother who is better at taking care of her baby's needs is more likely to survive, and therefore that mother's genes (including any associated with quick response to a needy baby's cry) will on average be better represented in the next generation. That principle repeated for a thousand or a million generations will ensure what common sense already tells us: Mothers are powerfully primed to respond with attention and nurturing to their babies' needs. Kindness to one's genetic relatives is, from an evolutionary perspective, hardly distinguishable from kindness to one's genes (or to a certain percentage of those genes). Kindness to strangers, however, is a major puzzle, both in common sense and in evolutionary terms[3]; and it is this glorious bit of exoticism, this puzzling and remarkable principle of kindness to strangers—or *altruism*, as it's sometimes called—that I want to explore further.

While herding camels, a Kenyan ranch worker came upon a group of elephants. The matriarch of the group, perhaps threatened by the appearance of this man, charged him and knocked him down with her trunk. His leg broken by the attack, the man was unable to move. A group of trackers found him the next morning, propped up against a tree but defended by a large elephant, who wouldn't allow them to approach. The trackers finally gave up, returned to the ranch, and the ranch manager drove out with a gun. As he approached and raised his gun, the injured man shouted out, told him not to shoot. The manager thus fired warning shots, forcing the elephant to retreat far enough that he could drive up to the worker, who then explained what had happened. Yes, the elephant had indeed knocked him down and broken his leg, but then, using her trunk, she pulled him over to the tree, where there was some shade, and gently pushed him upright, leaning against the tree, with her foot. Even as the rest of her family group moved away, the elephant stayed with him the rest of the day and for the entire night. She regularly touched him with her trunk, as if to comfort him, and once, when a herd of wild buffalo wandered nearby, she chased them away. The injured man believed that the elephant had deliberately tried to care for and protect him.[4]

A second story, told by primatologist Frans de Waal in his book *Our Inner Ape* (2005), features a bonobo named Kuni kept in captivity at the Twycross Zoo in England. One day Kuni watched a starling fly inside his enclosure, accidentally collide into a clear pane of glass, and drop to the ground on the inside of the moat separating Kuni from the human public. Kuni went over, picked up the stunned creature, then carefully tried to place him or her upright. The bird didn't respond. Kuni gave a toss, which resulted in some ineffectual fluttering. Then the ape picked up the bird once more and began climbing up to the top of the highest tree in her enclosure. Up there, high and swaying, holding on with both her feet, Kuni used her hands to grasp the bird gently. She opened up the bird's wings, then once again gave a toss—and the bird flapped and fluttered and then fell, finally, down to the inside edge of the moat. Kuni descended from the tree and stood guard over the unmoving creature, not allowing any of the other bonobos to approach, for some time. Finally, the starling recovered sufficiently to fly away. "The way Kuni handled this bird," de Waal remarks, "was unlike anything she would have done to aid another ape. Instead of following some hardwired pattern of behavior, she tailored her assistance to the specific situation of an animal totally different from herself."[5]

A third account involves a young female gorilla named Binti Jua, living in the Brookfield Zoo of Chicago. Zoo-goers were able to look down into Binti Jua's enclosure from above, protected as they were from the wild animals by an eighteen-foot wall: too high for any gorilla to climb but not so high that a three-year-old boy couldn't fall in, strike the concrete floor, and still survive. As the little boy's mother screamed at the horror of what had happened, Binti Jua went over to pick up the unconscious child. She gently held him, sitting down on a log in the exhibit and patting him on the back, then carrying him over to an access door and handing him off to a member of the zoo staff. The boy was saved by a gentle and sympathetic gorilla.[6]

You might be inclined to dismiss these three tales as mere anecdotes, but they have been selected as useful case studies by well-known scientists of animal behavior: elephant expert Joyce Poole, primatologist Frans de Waal, and cognitive ethologist Marc Bekoff. You might also object to the case of Binti Jua because it is well-known that she had previously been trained by the zoo staff in the techniques of proper mothering. In picking up the human child, one might therefore argue, she was robotically carrying out a few simple learned behaviors.

And, of course, none of these stories describes altruistic behavior toward a member of the same species—so consider the case of a chimpanzee named Jessie, living in a half-acre compound in the southern United States. Jessie was often observed going to the drinking fountain to fill her mouth with water, then taking that mouthful across the area, clambering up to the top of a high structure where her friend lay waiting, then pouring the drink into her friend's mouth and assuaging the friend's thirst. Jessie may have perceived that her friend was thirsty and brought her a drink as an act of kindness.[7]

To be sure, none of these tales shows an act of kindness requiring serious risk, so let me suggest the following case, once again of chimpanzees. Chimpanzees can't swim. Their bodies seem to be significantly denser than human bodies, and they can't even float. They're usually so terrified of water that a moat or river or lake is effective in controlling their movements. For that reason, the chimps living at a primate research project in Norman, Oklahoma, were sometimes let loose on a small island in the middle of a lake. As an added measure of control, the edge of the island was reinforced with an electric fence. One time, however, an inquisitive three-year-old female named Cindy found a hole in the fence. She promptly fell into the water, splashed, then sank below the surface. In response, nine-year-old Washoe dashed over the fence and, standing at the edge of the shore, reached down and pulled out the drowning youngster.

Did Washoe ever risk drowning herself?[8] Maybe and maybe not. But a male chimp at Lion Country Safari in Florida who once tried to retrieve an infant similarly fallen into a moat lost his own life during that heroic attempt.[9]

Altruism is the generous assistance to others in need, an act in which the altruist incurs a cost to provide a benefit to someone who is not a close relative. Thinking as evolutionary biologists, we might describe such costs and benefits in terms of reproductive potential. The Good Samaritan, in stopping to help the robbers' victim, was not merely losing time and spending extra money; he was also taking significant risks that the thieves would soon find, rob, and kill him. So we can calculate the cost as a real risk of lost life and lost reproductive opportunity, while the gain to the victim was of unknown reproductive years to his life.

This was not a case of mutualism or nepotism. Nor was it an example of reciprocal cooperation. The Good Samaritan did such things without an expectation of future return; and since he did so discreetly and in the dark of night, we can't even say that he hoped to benefit by improving his own reputation as a generous man.

Why should the altruist do such a thing? In the parable (or at least in the version I find in my modern English translation), we are told that the Good Samaritan did it because the sight of the helpless and injured victim caused him to have "compassion." Compassion was the Samaritan's strong response or psychological experience that made him behave altruistically. This biblical sort of compassion I will treat as synonymous with a term many scientists prefer: *empathy.*

Let me define *empathy* as "knowing resonantly what someone else is feeling." A person can know things, or know about things, intellectually and abstractly. But when I speak of *knowing resonantly,* I mean knowing as a form of experience, as when your feelings are responsive to, or sympathetic with, or in resonance with the feelings of someone else. To experience empathy, then, would mean you comprehend someone else's feelings because, to some degree, you have been provoked to experience them yourself. You know another person is sad because, somehow, the sight of her suffering makes you feel a little sad, too.

Because I've limited my definition of empathy to an emotional knowing, I wouldn't call understanding someone else's visual or auditory perspective a case of empathy. A rhesus monkey can follow the gaze of another monkey, or of a person, and acquire a sense of what that other individual is seeing. Rhesus monkeys also have a sense of what other individuals are hearing, and they understand the difference between seeing and hearing.[10] The ability to appreciate another's perceptual viewpoint is a fascinating aspect of human and animal intelligence, but it's not what I call empathy.

Nor will I insist that the experience of empathy always results in a behavior of altruism. It's common enough to think so, and since empathy can't directly be seen, one reason we believe it exists is that we witness the results in altruistic or helping behavior: an attempt to alleviate someone else's suffering. But clearly someone can feel empathy for someone else's distress and not try to reduce that distress. The Good Samaritan may not, after all, have been the only one who felt empathy. The other two travelers may well have had the same psychological experience, so we can imagine. The real difference, then, was that the other two didn't perform an altruistic act. Why not? Perhaps a reasonable fear of being attacked themselves by thieves overwhelmed any urge to kindness, and so, in spite of feeling compassionate—or empathetic—they crossed to the other side of the road and hurried on their way.

Another example of empathy without altruism: I might say that the chimpanzee Austin understood empathetically that his enclosure-mate, chimpanzee Sherman, was seriously afraid of the dark. But while routinely making a show of trying to comfort Sherman and reduce his fear, Austin simultaneously exploited it. Sherman was socially dominant to Austin, and during the daylight hours Sherman would bully Austin and in general expect him to behave as a chimpanzee subordinate is supposed to. Once it became dark, however, Austin was able to use Sherman's weakness to reverse the situation. After dark, Austin would casually wander out of their night quarters and deliberately make strange noises that frightened poor Sherman, huddling anxiously inside, and then Austin would rush back in to comfort his distressed superior, thereby temporarily gaining an advantage.[11]

Philosopher Daniel Dennett has suggested that one way to think about minds and intelligence is to consider orders of intentionality. Zero-order intentionality would describe an animal's mind as being devoid of any subjective experience. In first-order intentionality, the animal wants something—say, a peach—or possesses a mental representation of a peach, but without any sense of others as beings with minds. In second-order intentionality, the animal recognizes that another animal also has a subjective mental experience, such as a desire or plan or idea. Second-order intentionality, in short, describes mind considering mind.[12]

We might imagine that empathy requires second-order intentionality. Austin was aware that Sherman's mental experience included a fear of the dark.[13] And it's only too easy to conclude, based on the concept of second-order intentionality, that empathy is something rare that requires complex cognition or high-level learning over time. Indeed, all the examples of empathy and altruism I've given so far refer to a small elite of charismatic, large-brained mammals—elephants, gorillas, bonobos, chimpanzees—all well-known for being behaviorally complex and cognitively sophisticated. But, following the model recently promoted by neurologist Stephanie Preston and primatologist Frans de Waal, I would prefer to consider empathy as appearing in two different but related forms, *contagious* and *cognitive*, with the first—contagious empathy—a lot simpler and more widely distributed than we ordinarily expect.[14]

At its simplest, contagious empathy is the process whereby one person yawning in a midafternoon college classroom sets off a quiet cascade of yawns among the entire class. Or the process in which a single bird, startled by some sudden movement, takes off in alarm and is instantly joined by the entire flock. This kind of contagion is shown by a number of species, from American ground squirrels to olive baboons in Africa, and indeed, I suspect I see it in the antics of my two dogs, Smoke and Spike. When Smoke barks and starts chasing a shadow at dusk, Spike automatically barks and leaps forth to join in the chase.

Contagious empathy happens automatically, often well below or beyond the sweeping light of rational thought and conscious awareness. Human infants barely out of the delivery room, at the age of one and a half days, will automatically cry in response to the distress cries of other babies; and they immediately distinguish between a baby's cry and a computerized simulation of one.[15] That's contagious empathy as it appears among *Homo sapiens,* and it seems remarkably similar to what we see in other mammals. Mice and rats, for example.

We know that mice and rats feel a significant stress (as indicated by elevated blood pressure and pulse rates) when they watch laboratory researchers decapitate other mice and rats.[16] That may be contagious empathy.

We also know that mice react to pain felt by other mice. When you inject a mouse with a 0.8 percent solution of acetic acid, you cause a painful sensation marked by rhythmic abdominal constrictions, or writhing. The writhing can be measured, and so it gives researchers a convenient way of identifying the intensity of a mouse's pain. Using a writhing count as their measuring device, scientists at McGill University recently placed mice in pairs inside transparent Plexiglas cylinders, and then they injected one of each pair with a particular amount of acetic acid solution. The injected mouse who had been paired with another mouse writhed with a significantly greater intensity than a mouse injected with the same amount of acetic acid but kept in isolation. Moreoever, when the McGill scientists injected both mice in a pair, they found that each mouse writhed more intensely than a mouse injected with the same amount but paired with a noninjected mouse. "These findings imply," the researchers conclude, "the communication of pain from one mouse to the other."[17]

Was this merely a case of simple imitation? Probably not, since the researchers also found that an acetic-acid-induced pain in one mouse would affect the responsiveness to heat-induced pain in a second mouse paired with the first. The empathetic communication of pain, moreover, increased if the paired mice knew each other (that is, had been cagemates for a particular number of days). Inject two cagemates with the acetic acid solution, and they will each show significantly greater writhing than two mice in the same circumstances who are strangers to each other. Additional experiments showed that the only way to reduce this communication of pain was to place a visual barrier between the paired mice. In other words, the contagion spread through vision—watching another mouse in pain—rather than through hearing or smell.[18]

The ability to feel pain is an obvious example of evolutionary continuity. It's a neurological, physical, and psychological feature with broad distribution across a great zoological swath. Pain engages a part of the brain called the *anterior cingulate cortex*, which processes the sensory input and introduces a coherent mental-and-physical experience that is often expressed automatically with a cry, a scream, a yelp—whether you're a person or a dog or a nocturnal prosimian living on a tiny island off the coast of Madagascar. Such is evolutionary continuity; and a growing number of scientists believe that contagious empathy, the automatic ability to feel someone else's pain, is also evolutionarily continuous.[19]

Evolutionary continuity in a psychological experience implies neurological continuity, and the neurological mechanism involved in empathy could be associated with specialized brain cells known as *mirror neurons*, which I referred to briefly in an earlier chapter. Mirror neurons were discovered and named by Giacomo Rizzolatti and his colleagues little more than a decade ago. The initial discovery began as an accident that happened in a neuroscience laboratory at the University of Parma, in Italy, while Rizzolatti and his colleagues were mapping out the function of cells in a brain region (identified as F5) in the premotor cortex of rhesus monkeys. This part of the monkey's brain is associated with grasping and moving a seen object. When a laboratory monkey saw and moved his hand to a piece of food and grasped it, the neurons in area F5 would fire. The researchers could observe this neurological activity because they had implanted electrodes into that region of the brain, and thus neural excitations would be registered on an external monitor. The monkey grabs a peanut, brings peanut to mouth, and the monitor indicates that neurons in region F5 are firing.

Then one hot day, a researcher in the laboratory walked in grasping and licking an ice cream cone, and the same neurons in the monkey's brain started firing.

It was astonishing. The monkey was merely watching someone else grasp and move an object to his mouth.[20] When the researchers tried the same thing with nuts and bananas and raisins, they got the same results. Those F5 neurons were activated both when the monkey grasped the nuts or bananas or raisins and drew them toward his mouth, and when the monkey watched someone else do it. This general principle was true whether the monkey was watching the grasping hand of a person or another monkey. It was true whether the seen object was near or far away. And it was true whether the object was food or not food. Nor was there any relationship between this neurological finding and any reward. The laboratory monkeys would respond the same whether or not they got to eat the banana or peanut or raisin in the end.

When the researchers began studying another part of region F5, they found neurons that were specifically related to actions of the mouth. About one quarter of them had mirror properties, with some of the mirror neurons responding to eating activities of the mouth and others responding to communication actions (smacking lips, for example). Later examinations of another part of the brain, the superior temporal sulcus (STS), found neurons that will respond to other body actions, such as walking, arm movements, head turning, and torso bending—done by the monkey and seen by the monkey when others do it.[21] Monkey see and monkey do were neurologically the same thing.

"It took us several years to believe what we were seeing," Rizzolatti recently told a reporter for the *New York Times*.[22] But what they were observing was less remarkable than what the presence of mirror neurons seemed to imply. Mirror neurons appear to enable some animals (including humans) to learn through imitation. I watch someone do something, and I can acquire enough knowledge strictly from that observation to do it myself. Among people at least, this uncanny ability to imitate others is present from the very start. Babies in the delivery room stick out their tongues in response to tongue-waggling grown-ups.

Mirror neurons have a second important function, which is to give us a quick way to understand and predict the actions of others. Whenever we see a person grasp a coffee cup in one particular fashion and drink from it, we review a vivid neural representation of the act that includes both the visual and the motor aspects. Then, next time we see someone grasp a similar cup in a similar fashion, we have the knowledge to predict quickly and accurately that his or her intention is to drink from it, rather than, say, throw it at us. This quick prediction of someone else's goal in an action would be useful in an evolutionary sense: Imagine a rock instead instead of a cup.

But mirror neurons may also improve our understanding of social behavior more generally, since they give us a powerful way to understand other people's emotions, in part, perhaps, from witnessing the visual display of emotions on the face. In essence, we transform the experience of seeing someone else have an emotion—experiencing a humiliating rejection, for instance—into experiencing it ourselves. We can be entertained by televised sports because we are able not only to see but to experience mentally, through activated neurons, the motor skills involved in athletic performance; and we are entertained by film and theater because we can experience emotionally the emotional lives of others. We feel this woman's magnificent triumph, just as we experience that man's humiliating rejection. Mirror neuron systems may, in short, provide an internal simulation device by which one person can appreciate and imitate the actions, perceptions, and emotions of another; thus, contagious empathy may be grounded in our own bodily experience at the neurological level.[23]

Cognitive empathy is contagious empathy pressed through a cognitive filter: a brain or mind. Cognitive empathy, in other words, is a developed or elaborated version of contagious empathy, an experience or inclination that has been affected by different levels of self-awareness, by competing emotions, by reasonable assessments about possible cost and outcome, and so on.

Contagious empathy is the simple, resonant communication of pain or distress. One distressed and crying day-old infant makes another day-old infant distressed and crying. One mouse in pain increases a fellow mouse's pain. In cognitive empathy, by contrast, we see the evidence of other emotions, ideas, and capacities at work. Day-old infants are cognitively unprepared for identifying a separation between themselves and others. Babies who are eighteen months old, however, can make that distinction; and when they see another crying baby, they experience the distress but they also seem to appreciate that it is the distress of someone other than themselves, and they will often try to alleviate the other baby's distress.[24]

Life is not so different for rats. A classic demonstration of cognitive empathy among rats is the hoisting experiment, originally reported in 1959. In this experiment, researchers caused albino laboratory rats to feel uncomfortable or distressed by wrapping them in tiny, corset-style harnesses with their legs free to move and then hoisting them off the ground, using a small, Erector-set motor to run the hoist. The rats in the hoisting compartment would usually squeal and wriggle madly once they were suspended, and if a rat was not squealing and wriggling to the satisfaction of the researchers, "it was prodded with a sharp pencil until it exhibited signs of discomfort."[25] Twenty experimental rats were then, one by one, placed in an adjacent compartment and allowed to watch the distressed rat through a transparent window for ten minutes, while having the option of lowering the creature by pressing a bar. Even with no training in working this peculiar apparatus, the watching rats regularly pressed the bar, lowering the distressed rat back down to the floor of the compartment. For comparison, another twenty rats were put in the same situation—adjacent compartment with bar—except that instead of a suspended fellow rat, they watched a suspended block of Styrofoam. They occasionally pressed on the bar as well, but at a significantly reduced rate.[26] In short, watching a fellow rat in distress produced an experience of cognitive empathy that provoked the watching rats into carrying out rat-size acts of kindness for the benefit of the distressed rats.

Interestingly enough, a later parallel experiment, in which a rat in one compartment watched a distressed rat in an adjacent compartment, produced very different results. In this case, however, the other rat was distressed by genuine pain caused by a continous electrical shock coming from an electrified floor grill. Once again, the watching rats were given the ability to turn off the source of distress—this time, the electrical shock—by pressing down on a bar. Now, however, the watching rats turned out to be significantly less likely to press down the bar when the observed rats were being shocked than when they were not being shocked. It was an apparent reversal in helping behavior, and the reason for the reversal, so the researchers concluded, was that the watching rats were, in fact, overwhelmed by a fear response to the sight of a fellow rat being shocked. As the researchers noted, the watching rats "typically retreated to the corner of their box farthest from the distressed, squeaking, and dancing animal and crouched there, motionless, for the greater part of this condition."[27] In short, a contagious empathy, experiencing another rat's pain, rather than being transformed into an urge to help, was affected by an overwhelming fear that produced the urge to avoid—rather like the two men in the Good Samaritan story who, upon seeing the victim of a robbers' recent attack, moved to the other side of the road and hurried along.

Still, the fear produced by watching someone else get shocked need not be so overwhelming, as we can conclude from the results of another experiment, this one with rhesus monkeys. Fifteen rhesus monkeys were taught to acquire food treats by pulling on one of two chains, either of which would deliver the same amount of food. The monkeys were also able to look through one-way glass into another compartment where another monkey was feeding. The other compartment, however, had been lined with an electric grid, and on the fourth day of the experiment, the grid was activated so that whenever the experimental monkeys pulled one of the two chains, the watched monkey would receive a shock. Pulling on the second chain delivered no shock.

After the shocker was activated, two thirds of the monkeys quickly developed a significant preference for pulling on the chain that did not shock the other monkey, and of the one third of the group who showed no preference, two actually stopped pulling on either chain altogether, one for five days, the other for twelve. Thus, ten of the fifteen monkeys chose less food—hunger—and an additional two chose no food—starvation—rather than cause pain to a fellow monkey. Only three of the monkeys showed no particular reaction whatsoever. Further research demonstrated that the important effect was visual—seeing another monkey being shocked—rather than auditory. The monkeys also were somewhat more inclined to avoid shocking a monkey they knew, a cagemate, as compared to a stranger.[28]

Increasingly, such laboratory experiments on the empathy and helping behavior of mice, rats, and monkeys are being supplemented by field studies of animals big and small. Big takes us up as far as elephants.[29] Small brings us down to bats. Watching a nesting colony of fruit bats in Florida, biologist Thomas Kunz observed that a pregnant female, about to give birth, was still suspended by her feet, with her head down. Such is the usual position for a bat at rest, but to give birth, females are required to reverse that position, with their heads up and feet down. To the biologist's astonishment, however, another female, unrelated to the first, landed next to the pregnant one and assumed the correct birthing position. It looked like a demonstration, Kunz thought, and indeed the pregnant female seemed to respond to the modeling. She reversed position, ready now to give birth. Next, the helping bat began licking the mother's perineum, as if to promote the physiological process of birth, and when the baby bat began to appear, the helper stroked and groomed him, then seemed to guide him as the tiny infant blindly moved toward the mother's lactating breasts.[30]

Empathy, both contagious and cognitive, is the tie that binds. Empathy is what caused a traveling Samaritan to become the Good Samaritan. Empathy, at its extreme, is the origin of self-sacrificing heroism. But self-sacrificing heroism is rare, and so we might best think of empathy, and the altruism it promotes, in their more common and mundane versions: as daily experiences of concern and small acts of kindness.

An action requires two things to qualify as altruism. First, it must cost the actor while benefiting the recipient. Second, the actor and recipient cannot be close genetic relatives. That they can't be close relatives makes intuitive sense, given that we all understand the enduring connection of close relatives: the emotional ties of brotherhood and sisterhood, for example, that can produce acts of self-sacrifice requiring little explanation.

That proviso is also important for thinking in the style of evolutionary biologists. In considering behavior from an evolutionary point of view, we're speaking in terms of reproductive success: the movement of genes from one generation to the next. If the altruist incurs a genuine cost—risking his or her genetic representation in the next generation—it can be hard to see how any inherent inclination to altruism will be maintained through evolution. Altruism, in short, seems like a loser's strategy. Oh, yes, it's admirable. It's inspiring to see. We give medals and parades in honor of the woman who sacrificed her life so that another human being, a perfect stranger, might live. But the woman who sacrificed her life has also sacrificed her genetic future. Any genes she possessed that happen to be associated with such self-sacrificial behavior have not moved into the next generation. Wouldn't, then, the infinitely patient winnowing of evolution simply cast away all genes associated with altruistic self-sacrifice? And why should such genes be there in the first place?

If we can imagine a gene having a goal, then we can say that a gene's single goal is to replicate itself. If it fails to do so, it ceases to exist and will thus be replaced in the universe by genes that are better at replicating themselves. In this simple and single-minded competition with other genes, we can imagine a gene as being "selfish"—and that evocative metaphor is the basis for the selfish gene theory, a way of thinking about evolution popularized by Richard Dawkins in his book *The Selfish Gene* (1976). Some people believe, in error, that the selfish gene theory means evolution cannot possibly transmit unselfish altruism, that the blind and endless competition among genes inevitably means that evolution can operate only through a blind and endless competition among individuals. Dawkins himself made that mistake. As he wrote in the opening salvo of his bestselling book:

> Be warned that if you wish, as I do, to build a society in which individuals cooperate generously and unselfishly towards a common good, you can expect little help from biological nature. Let us try to teach generosity and altruism, because we are born selfish. Let us understand what our own selfish genes are up to, because we may then at least have the chance to upset their designs, something that no other species has ever aspired to.[31]

It's a bleak vision, warmed only by the kindness of his warning—and cooled again, once we recognize the selfishness inherent in a message that would make his book required reading. It also fails to explain why inherently selfish humans should ever have any interest in hoping to build a better society where humans act unselfishly. Those who hold this belief often go on to insist, as Dawkins does, that true altruism is a behavior unique to humans, and since it cannot be transmitted by biological evolution, by default it demonstrates the workings of divine or—as Dawkins would have it—cultural intervention in human affairs.

In fact, however, there is no sound reason to imagine that inevitable competition at the genetic level means evolution must always proceed through open competition among individual organisms. Let's return to the example of the woman who lost her life in an attempt at rescuing someone else, a stranger she randomly discovered to be in serious trouble. In sacrificing her life, she also cut short her reproductive career. Too bad for her genes. But her premature death may not at all represent the usual or average consquence of having her particular package of empathetic and altruistic genes. It could be that for every heroic man or woman who has died trying to rescue someone else, another dozen equally altruistic individuals have survived and found their social and economic status usefully elevated as a result of such actions. In addition, the very genes associated with high risk-taking and high levels of compassion might be simultaneously associated with more common behaviors that make a person engaging and dramatic—and as a consequence reproductively more successful than average. The altruistic woman died prematurely, a failure from her genes' point of view, but maybe her case is an exception. Maybe, on average, that same genetic packet is very successfully moving into future generations.

That's a general way of thinking about it. Maybe too general. But the vision of altruism and empathy I've been developing in this chapter, based significantly on the work of Stephanie Preston and Frans de Waal, provides a more particular way of explaining both the act and its motivation. The Preston and de Waal model avoids holding up altruism (or "helping behavior") and empathy as if they were singular exotic entities to be dissected in isolation. Instead, the model would combine helping behavior with contagious and cognitive empathy, and a few similar phenomena, into a unified whole. The combination makes a unified whole because all the elements are, or could be, productions of a single kind of empathetic neural system, perhaps the system associated with mirror neurons.[32]

Seeing the problem this way changes the nature of the question. One no longer asks, how did empathy or altruism evolve? Instead, one asks, how did the neural system that supports empathy, altruism, contagion, and similar features evolve? Once we see empathy and altruism as parts of a larger whole, we are also prepared to recognize the considerable benefits that accrue to that full system. For example, the full system makes mothers instantly and automatically responsive to the needs and actions of their infants. It allows many members of a social group to benefit from the observations of single individuals, essentially transforming a group of ten separate animals, each with only two eyes, into a coherent living entity with twenty eyes. It helps coordinate group actions quickly and automatically, thereby improving any individual's chances of survival from predators and increasing an individual's harvest of prey.[33]

As a daily event in the lives of humans, empathy emerges as a continuous series of small and unexpected acts of kindness done without expectation of recompense. In these small actions, empathy serves its most important social function, as an expressive, positive, cohesive connection between individuals. Empathy is like the monkey-rope described in Herman Melville's novel *Moby-Dick*, a social safety line that ties a stable person on board the ship to an unstable one working off the ship's side, that man wielding a sharp butchering spade and precariously perched on the slippery crown of a whale's carcass. A single slip, and he will slide helplessly down and be crushed in the ever-shifting gap between a giant's corpse and the ship's hull. Worse, he could fall the other way, down into the icy sea, and join a snapping feast of hungry sharks. The monkey-rope keeps a precarious individual secure, attached to the stability of his partner and the promise of a living society aboard the ship.

As Melville develops this idea, we see that the generous connection between friends or strangers, this monkey-rope, resembles the natural connection between siblings. Twins, even. ("Queequeg was my own inseparable twin brother.") It's a marvelous image that turns also, in its metaphorical elaboration, mysterious. How is it possible for two strangers to be so connected that they will treat each other in the way of brothers? The answer takes us back to the origins of altruism in the experience of empathy, and from there it takes us to the evolution of a neural system that predates human evolution by millions or tens of millions of years: a system, therefore, that we might share with a significant number of other social species.

PART IV

Where Is Morality Going? Assessments

CHAPTER 12

Duality

> Another point of difference between the male and female schools is still more characteristic of the sexes. Say you strike a Forty-barrel-bull—poor devil! all his comrades quit him. But strike a member of the harem school, and her companions swim around her with every token of concern, sometimes lingering so near her and so long, as themselves to fall a prey.
>
> —Herman Melville, *Moby-Dick*[1]

Rules versus attachments: To think about the difference between the two, we might first imagine rules as the origin of *right* and attachments as the source of *good*. Rules, after all, appear to tell particularly what is allowed and what is forbidden, the literally right versus the literally wrong. I say *literally* because even while the rules may be psychologically embedded and socially enforced, in human societies the rules are also written down, and people will return to the written rules again and again for consultation about specific situations. Attachments, however, may be closer to intuitively felt general principles, an inherent and generalized sense of what must be good.

Often right and good amount to the same thing—but sometimes not. Consider, for example, the well-known biblical story (eighth chapter of John) in which a group of scribes and Pharisees brings an adulterous woman into the temple where Jesus has been teaching. The scribes and Pharisees, scheming to mark Jesus as a false prophet or a demagogue, remind him that the law according to Moses would punish the adulterous woman with death by stoning. And then they ask him, "What do you say about her?" Jesus wisely avoids overtly contradicting the law or rules, while simultaneously he appeals to his listeners' sense of kindness and fairness. He says, "Let him who is without sin among you be the first to throw a stone."

Here's another instance, taken almost randomly from the contemporary news, of a conflict between rules or attachments, or right versus good. In Sudan, thirteen women were arrested by the police for wearing trousers outdoors, a sartorial act done in flagrant violation of the nation's morality laws, which are based on a strict interpretation of Islam. Ten of the women accepted their punishment, which was flogging combined with fines, while three chose to take their case to a public trial. Among those three was an outspoken journalist, Lubna Hussein, who declared that she would rather stay in jail for a month and receive forty thousand lashes than submit to such laws. "I will not pay a penny," she told a reporter for the Associated Press, insisting that such morality laws were not only bad themselves but actually violated the Sudanese constitution. "When I think of my trial," Hussein said, "I pray that my daughters will never live in fear of these police. . . . We will only be secure once the police protect us and these laws are repealed."[2] It might be right, according to the law, to punish this woman, but it would certainly not be good.

You could say that both examples I've just given are extreme cases, or you might argue that the rules cited—the law of Moses or the Sudanese morality laws based on a strict interpretation of Islam—are misinterpretations of the real rules and not, therefore, valid instances of rules morality. So perhaps we need to consider case studies that are more familiar and closer to home. The more familiar and closer-to-home examples, however, can be harder to appreciate or parse. You may already have firmly entrenched opinions about the moral issues involved in, say, the current debate in the United States over a woman's ability to control what happens to her body versus the rules supporting sanctity of life, or the attempts to legalize gay marriage as an act of fairness versus the various passionate references to biblical authority on the subject of marriage and family structure.

Sometimes the appeal to rules morality is easy to identify, as when a person refers to some specific code or text—quoting one of the Ten Commandments, for instance—that seems to pronounce so unambiguously what's right and what's wrong. But sometimes a reference to attachments morality may be disguised as, or confused with, a reference to rules morality: for example, the recitation of established "rights," such as human rights, civil rights, women's rights, animal rights, and so on. I generally tend to agree that thinking in terms of such "rights" is useful, but while they may seem like appeals to universal moral rules, they are actually appeals to universal moral attachments. The idea of animal rights has recently been invented in imitation of civil and human rights, but it operates mainly through stimulating our empathy for suffering or neglected animals. Similarly, civil and human rights have been established in codes and written into law as if they are a matter of what is right, but they are still based on what we feel and know emotionally—*in our hearts*, as we often say—to be good.

As if to complicate things even further, these two sometimes contesting entities, rules and attachments, are invariably colored by gender, with rules morality often associated with men and attachments morality more frequently associated with women.

Indeed, one of the abiding curiosities of life—of human life—is that the moral rules are usually owned by men and too frequently endured by women. By *owned*, I mean created, defined, interpreted, and enforced. We see the archetypal pattern in the story of Moses, a man who, having received the law from God (Himself imagined and described as masculine), presented that law to his cadre of exclusively male judges and an exclusively male priesthood. The very language of the Ten Commandments indicates that they were addressed to men only, and the law of God has traditionally, with only a few exceptions even in our enlightened present, been handled and interpreted by an exclusively male line of imams, rabbis, priests, and preachers across the spectrum of monotheistic traditions.

If this were true only of monotheism, if we could find a solid sample of alternative religions and cultures where women comparably owned the rules, we might then attribute the human male ownership of moral rules to some strange historical accident, a random case of cultural construction that, like a meteor dropping out of the sky, *just happened*. But, you would be hard-pressed to come up with a clear instance of any human culture, present or past, where the rules are comparably owned by women. Yes, people have fervently longed for and even brightly imagined such cultures. In Melville's time, Johann Jakob Bachofen, a German attorney, famously promoted his theory of *Das Mutterrecht*, or *The Mother Right* (1861), which declared the rules and laws to have originally been established by women, an historic act that brought all of humankind out of the barbaric shadows and into the light of civilization.[3] Friedrich Engels expanded on Bachofen's theory with the idea that this early matriarchal period in human history was superseded once the arrival of agriculture and the domestication of animals led to settled communities and private property, which were then monopolized by men.[4] It's an interesting speculation, perhaps, but nothing more. No anthropologist has ever seen such an Edenic human culture run by Eve, nor has any archaeologist ever discovered evidence of its past existence.[5]

Certainly, we can recognize that human cultures may significantly differ in their treatment of women and in the integration of women into political life. And what about those famously egalitarian hunters and gatherers? Don't they provide the historic instances where the rules have been equally owned by men and women, in a simpler and more satisfying world of shared goods and shared power?

Elizabeth Marshall Thomas wrote the classic account of life as she experienced it among the !Kung (or Bushmen) of the Kalahari Desert of Botswana, in her book *The Harmless People* (1959).[6] The word *harmless* in the title, taken by many anthropologists to indicate these were a remarkably peaceful people who seldom or never killed anyone, was meant by Thomas more simply as a translation of how they describe themselves: *Ju/hoan* or *Ju/wa*, meaning the people who are "clear" or "pure" or "not armed"—or "harmless."[7] The !Kung, with a murder rate higher than that found in any American city, are certainly not more pacific than anyone else.[8] Later observers of the !Kung have portrayed them as an attractively egalitarian people, where—in the words of anthropologist Richard Lee—there is no evidence that the women are "oppressed or dominated" or "subject to sexual exploitation."[9] Nevertheless, Lee found that men more often become spokespersons of a group, and that two thirds of any public discourse is carried out in the voices of men.[10]

Anthropologist Marjorie Shostak has similarly found the !Kung to be nearly egalitarian, and in her assessment, mothers have roughly the same authority as fathers in a family. Altogether, Shostak reports, women among the !Kung show a "striking degree of autonomy."[11] But Shostak also provides a more direct portrait of life among the !Kung, in her transcription of the spoken autobiography *Nisa: The Life and Words of a !Kung Woman* (1981). Here we learn that Nisa was regularly beaten by her husband, who would fall into jealous rages, once striking her brutally all over her body with a sharpened stick, another time cutting her with a knife and nearly killing her. Nisa's daughter, Nai, was similarly subjected to her husband's abuse. Nai was taken in marriage by an older man before she had reached puberty, and after she reached adolescence, he insisted on having sex during her first menstruation, even as she refused. Her refusal so enraged the husband that he knocked her down hard enough to break her neck. She was killed, and so her mother, Nisa, appealed to the tribal headman for justice. After a meeting of elders, the headman delivered the verdict, which was based on his and their understanding of the rules. "You fool," the headman said to the husband. "When a young girl has her first menstruation, you don't have sex with her. You wait until she has finished with it. Nai knew what she was doing when she refused you. Yet, you went ahead and killed her!" The husband was ordered to make restitution—five goats—to Nisa, the mother of the dead girl.[12]

In short, even among the supposedly "peaceful" and famously "egalitarian" hunter-gatherers of the Kalahari, living in conditions of extreme economic simplicity, it would appear that rule interpretations and judgments are more often delivered by men. Men still seem to own the rules. Elsewhere, the pattern is numbingly similar.[13]

Why should this be? Why shouldn't men and women own the moral rules equally everywhere, in every culture? The traditional Judeo-Christian story of a God who was male by cosmic coincidence and, through a natural sympathy for his own sex, began the world's creation with a man and passed down his authority to men is unconvincing and unsatisfactory. Nor was early twentieth-century psychoanalysis much more helpful in explaining such an odd asymmetry. Sigmund Freud, the inventor of psychoanalysis, held that morality emerged with the development of a superego, which, so he imagined, amounted to the mind's repository of parental values and rules. But the superego was formed as a result of experiencing an Oedipus conflict and castration anxiety, and since young girls did not go through an Oedipus conflict and had no good reason to be anxious about castration, Freud could only weakly conclude that they remained inferior to men in their moral understanding. Women had "less of a sense of justice," he wrote, and were "more often influenced in their judgements by feelings of affection or hostility."[14]

Freud's early work on the subject now looks naive and simplistic, but Swiss psychologist Jean Piaget, who, in the mid-twentieth century, more methodically studied moral development by observing children playing games, arrived at a similar conclusion. Piaget found that boys were increasingly attracted to rules and the resolution of conflict through debate and analysis of the rules, while girls seemed simply less moved by the rules altogether. They were more pragmatic and more willing to drop or modify rules in response to novel situations. Since Piaget presumed that the rules of game playing were perfectly analogous to legal and moral rules, he concluded that girls showed less of a legal acuteness, and that the moral sense was "far less developed in little girls than in boys."[15]

If we subscribe to the theories of Freud and other psychoanalysts, or to the later social psychology of Piaget and others, we might conclude that, as a result of some arcane developments in the womb or early childhood, or because of a socialization process taking place in middle childhood, girls mature more slowly and less completely into a moral understanding of the world than boys do. That's one way of looking at it. But later researchers would see a different pattern and arrive at some very different conclusions. In two studies published during the mid-1970s, Norma Haan and Constance Holstein argued that women's morality was not less developed but simply different from men's, and that it tended to rely more on an underlying sense of empathy or compassion.[16] Women's morality, they thus concluded, had a different tone or style from men's . . . or a different voice, as the idea was more famously articulated a few years later by psychologist Carol Gilligan, in her bestseller *In a Different Voice* (1982).

Gilligan found the young women in her own interview-based research to be hesitant, less certain of their moral analyses than men. "Women have traditionally deferred to the judgment of men," she wrote, "although often while intimating a sensibility of their own which is at variance with that judgment."[17] When Gilligan listened closely to those voices, she discovered that women were more likely to consider moral problems in terms of "care and responsibility in relationships," rather than with the more typically masculine examination of "rights and rules."[18] A morality based on rules alone, Gilligan thought, was incomplete and likely to become oppressive and damaging. Referring to a pair of well-known biblical stories, she illustrated her point by comparing the moral styles of Abraham and the nameless mother who appeared in court before King Solomon, hoping to reclaim her child, who had been stolen and claimed by another woman. Abraham showed his full obedience to the rule of God by agreeing to sacrifice his own son in worship. But as the wise King Solomon understood, no mother would likewise sacrifice a child merely to sustain a rule or principle, and thus he was able to distinguish the true mother from a false one by threatening to divide the child in half.[19]

Translated into sixteen foreign languages and selling nearly a million copies, Gilligan's book has become a feminist classic, while *Time* magazine anointed her one of America's twenty-five most influential persons for 1995. However, her work has been challenged from a couple of different directions. Some have questioned the quality of her data. Others have been concerned by the work's possible implications.

Superficially, at least, Gilligan's idea of two different moral voices reinforces some of the standard stereotypes about fixed gender differences, and in doing so it could undermine an axiom of traditional feminism: that gender is an arbitrary social construction completely unrelated to one's biological sex. Many feminists have traditionally resisted the notion of biologically based psychological differences in gender altogether. After all, the idea that men and women are psychologically different in essential ways has historically been used, by men, to suppress women by keeping them "in their place." In the kitchen, for example, and out of the workplace. Gilligan, while promoting her fascinating theories and observations about gender differences in moral reasoning, simultaneously recoiled from any extended debate about the origins of such differences. "I find the question of whether gender differences are biologically determined or socially constructed to be deeply disturbing," she protests in a 1993 introduction to her book.[20]

Disturbing or not, the question remains, and so does the larger concern about whether her famous book unhappily resuscitates dying notions and damaging stereotypes. In speaking of two different styles or approaches to morality, each identified more fully with one sex over the other, aren't we just perpetuating a set of harmful stereotypes about human gender?[21]

We usually think of stereotypes as unfair oversimplifications applied meanly. But if we can think of them as potentially useful generalizations applied kindly, then I would like to explore the matter of difference stereotypes a little further—but focusing more broadly on mammalian rather than merely human gender. Elephants are an interesting example if only because males and females live in separate societies that rather cleanly demonstrate some behavioral and temperamental gender differences for all three elephant species (Asian, African savanna, and African forest elephants).

The sexual segregation happens because once males reach adolescence, they leave their natal families, while the females remain. That system reduces the likelihood of incest and inbreeding, and it works because the elephants have evolved to behave in that fashion. The males probably want to leave. Their mothers and the other females may be glad to see them go. But the temperamental differences between males and females are evident almost from birth, according to the pioneering studies of African savanna elephants done in Kenya's Amboseli National Park by Cynthia Moss and Joyce Poole. Even during their first year, males play more roughly than their female counterparts, and by the age of around four years, the males show an increasing tendency to move independently of their mothers, and generally to become more of a nuisance, pushing and bullying others, for example. In contrast, the female calves continue to stay close to their mothers and soon become involved in babysitting—that is, playing with and caring for the younger calves in their family group. At the start of adolescence at around fourteen, the males drift away or are driven out completely, and they join a bachelor group or wander by themselves.

The adolescent males are still not as big as a fully adult female, and not even half the size of the largest males. And they are still discovering what it means to be a male. They form bonded relationships, *friendships* if you will, with other males of their general size and age, and they can develop a friendly, imitative relationship—perhaps it's fair to think of this as a mentoring relationship—with the big bulls. "As the older males go about their business," Joyce Poole writes, "searching for estrous females, mating and fighting, it is entertaining to observe teenage males tagging along behind, watching their mentors. In a subordinate head-low posture, they will stand near or follow an older male, investigating each spot of urine that the older male has sniffed. The older males are gentle with these youngsters since, of course, at this age they pose no competition."[22]

As with the males, females form dominance relationships with each other, which means that some females predictably dominate others. But female dominance is decided largely by natural circumstance, with little open competition required. The most powerful female, the matriach, is simply the oldest and most experienced individual, the one who retains the fullest memory of both social (other groups and important individuals) and ecological (distant water holes, seasonal sources of food) circumstances. Elephant females, in short, live in a familiar world of family where social power, or dominance, is defined largely by circumstance and consensus, while the males have entered a less familiar world of strangers, mostly, where social rank will finally be defined rather rigidly by an individual's size, power, and testosterone level.

Unlike females, the males continue to grow for most of their lives, which makes the older males, into their forties and fifties, simply enormous. The biggest of them (for the African savanna species) will measure up to thirteen feet at the shoulders and weigh around six tons, and their size alone—twice that of the biggest adult female or a twenty-year-old male—is enough to ensure a proper deference from the others. Just in case anyone should underestimate their great size and power, the big bulls like to show off, to exercise their mighty, mighty might by snatching up bushes and knocking over trees—up to fifteen hundred trees per bull per year, according to one study in Zimbabwe.[23]

Testosterone is the wild card here, since the males periodically undergo radical surges in those male hormones during the *musth* phase that can transform them from normal, easygoing creatures into raging and dangerous beasts. Dangerous, that is, to other males, even somewhat larger males, who under other circumstances would have the upper tusk. The big males do occasionally battle ferociously, earthshakingly, for access to an attractively fertile female. Yet most of the time they don't need to fight, since usually, during their long bachelor existence, they have already established a hierarchy of dominance. The males understand already, without having to risk their well-being or lives in a physical confrontation, who's at the top and who is not, even when the musth phase is factored in. The males understand the hierarchy because they have reviewed it and tested it with one another throughout their younger years, in previous confrontations and through the games of tusk-sparring and trunk-wrestling. Trunk-wrestling is as it sounds. Two young males face off, mutually entwine their trunks, and wrestle: pushing, pulling, each trying to throw the other off-balance. You know it's play because no one gets hurt, and, indeed, often no one is trying hard enough to show serious tension or real strain.

I don't want to create the impression, as stereotypes too often do, that everything can be simply explained in fixed terms, or that the differences between males and females are perfectly rigid and utterly stark. In considering gender differences among elephants, we should recognize them in the way we do for people, that you can describe a feminine or masculine temperament and style that may be acceptably accurate on the average but woefully inadequate or simply wrong when applied to any particular individual. Katy Payne, a bioacoustician known for her discovery of long-distance infrasonic communication among elephants, writes in her book *Silent Thunder: In the Presence of Elephants* (1998) of having once watched two male elephants standing together at midday in extreme heat in the desert of Etosha, a large national park in Namibia. One was a very big bull, the other much smaller and obviously much younger. As Payne watched, the little bull leaned over against the big bull, whereupon the big guy lifted his ear as if unfolding a parasol, a giant beach umbrella, which provided shade for his diminutive companion. The two stood that way, sleepily waiting out the hot overhead sun, "for a long time," Payne reports. The older bull gradually began to doze, and as he did, his ear slowly relaxed, sagged, and eventually drooped into the narrow space between the two, which startled the older one awake, whereupon he lifted his ear again and stretched it out once more into its parasol position. True, the big bull may have improved his own self-cooling efficiency by keeping the one ear lifted, but at the same time he was protecting the little bull from both heat and possible sunburn.

Payne also recalls a time that three colleagues in her Etosha research project spent a half hour watching three bull elephants. One of them was down on the ground, probably injured or sick, and the two others were trying everything they could to press their companion back into an upright stance.[24]

So the males do form strong emotional attachments. Nevertheless, I believe that Payne would be the first to agree that elephants, too, have gender-sensitive value systems that will recall Carol Gilligan's notion of two different moral voices among humans.

Actually, Payne first came to Africa to listen to the voices of elephants. Having spent time with elephants in a zoo, Payne had begun to suspect that those large animals were communicating with each other infrasonically, that is, with sounds that were of such a low frequency they fell beneath the range of human hearing. A few whale species do as well, and as Payne knew from her earlier years of studying the songs of whales, such low-frequency sounds can also, if powerful enough, travel for long distances through water and even through rocks and earth and other substances that are barriers to higher-frequency sounds. So she went to Africa carrying some sophisticated listening equipment that included tape recorders capable of registering the infrasonic environment. Using that equipment and working in Amboseli with Cynthia Moss and Joyce Poole, Payne began listening to elephant voices, which include a complex variety of snorts, rumbles, roars, bellows, screams, trumpets, along with "a long half-muffled, half-shrieking sound that was always associated with play." ("One day," Payne recalls, "we watched two calves running away from their mothers through a field of grass much taller than themselves with their trunks stuck out straight ahead of them, their tails straight up in the air and their ears flapping, play-trumpeting at top volume as they escaped!")[25]

Poole had previously identified twenty-six sonically complex vocalizations made by adult elephants, nineteen of which were produced only by females, four only by males, the remaining three by both. So females might be considered the more actively vocal sex, with nearly five times the vocabulary of males; but when Payne began analyzing the elephants' calls, she also realized that the males *only* made individual, solitary calls, whereas the females were often vocalizing communally, with overlapping and sometimes chorusing calls. The distinction between the voices of males and females was unmistakable. It was as if, she writes, the females would often begin a sentence with the collective pronoun *We*, while the males were communicating with others always as the solitary *I*.[26]

Why might evolution have produced different values and styles for the two elephant sexes? From an evolutionary perspective, both sexes are shaped, in body and mind, to achieve exactly the same goal, which is to reproduce as fully as possible. But because of different circumstances, each sex will be inclined to follow a different strategy to achieve that common goal.

Different circumstances. As mammals, female elephants are given the task of bearing their young and then feeding them, during their helpless infancy, through nursing. Because both parturition and lactation are nutritionally costly, evolution has seen to it that the females usually produce only one baby at a time, and that they remain infertile during the two years of gestation and an additional two years of lactation. Thus, a normal adult female's life is punctuated by extended periods of infertility, which severely limits her already finite lifetime reproductive potential. At most, she may be able to produce around a half dozen babies. Males go through yearly musth cycles, a few weeks or months each year when the testosterone surge makes them unusually aggressive and obsessed by sex, but even out of musth they are still fertile and carry a continuously replenished supply of millions of sperm cells. A male's lifetime reproductive potential, then, can be very high.

Different strategies. An elephant male can hope to achieve the greatest reproductive success compared to other males—that is, produce the largest number of healthy and reproducing offspring—simply by mating with as many fertile females as possible. His best strategy for becoming an evolutionary winner is quantitative. The males compete with one another and take their dominance hierachies so seriously ultimately because rank in the hierarchy determines who gets to mate with the fertile females during their times of maximum fertility. And the females? Given their limited lifetime reproductive potential, is there any point in their competing with each other at all? Yes, of course there is. Evolution works through competition, and the female winners will be the ones who best care for the offspring they have. Each elephant female can hope to produce only a half dozen babies in her lifetime, and most adult females can expect to have around that number. There will always be a male to provide the sperm, and she will routinely cycle into a fertile phase when the time and conditions are right. Hence, the critical distinction among females, the one that separates evolutionary winners from losers, is how many of those babies survive to a healthy adulthood and begin themselves reproducing. In other words, evolution should favor good mothering. The best elephant mother is probably going to be the one whose genes most fully move into the next generation, so that any genes associated with being a good mother—temperamental calmness under duress, for example—are selectively favored for generation after generation after generation.

That general distinction between the sexes is true for most mammals. Males are often shaped, in body and mind, for a quantitative strategy. Females are usually shaped for a qualitative one. But such broad generalities quickly shatter into a great variety of particulars at the species level. In some species (including humans) males may find a complementary or alternative route to reproductive success by following the qualitative strategy. For these species, fathering—nurturing the offspring—can become important. Meanwhile, the male quantitative strategy can itself play out in a surprising number of ways.

Yes, there is the old-fashioned physical competition that we see in elephants, which produces rank and a male hierarchy based on size and power and aggressiveness. The hierarchy is developed over long social experience among the males, and sometimes it's tested and reinforced by threats and high-noon showdowns, even head-bashing, tusk-slashing, all-out battles. We can recognize elephants as a species in which the old-fashioned physical competition is important simply by looking at sexual dimorphism ratios—that is, the ratio of size between male and female. One might think of females as maintaining an ecologically appropriate size, meaning the ideal size for individuals given their evolved relationship with food sources, predators, prey, and so on, and then we have to wonder why in so many mammalian species the males are bigger. Males have to deal with the same environmental problems, so their greater size could be an evolved consequence of males physically competing with other males, which will select for bigger males. Elephants males are twice the size of females, so we can reasonably conclude that physical competition among males is an important part of their reproductive style.

Males twice as big as females approaches the ordinary extreme for sexual dimophism, although sperm whales—Herman Melville's own study species—surpass that, with males growing to be roughly one and a half times the length of females and (at forty-five tons) three times the weight.[27] Sperm whales, incidentally, have a social system reminiscent of elephants', where the adult females and their offspring form their own social groups separate from the males. These are what Melville's narrator, Ishmael, describes as the "harem schools," and they'll typically consist of somewhere between ten and forty individuals. The females are sometimes related and sometimes not, but they share babysitting and possibly nursing duties. They are emotionally attached to one another, or affiliated, and will cooperate in defense against such fearsome predators as killer whales (orcas) and killer primates (human whalers). They will risk their lives defending others in their group,[28] so Melville's Ishmael was accurate when he stated, "But strike a member of the harem school, and her companions swim around her with every token of concern, sometimes lingering so near her and so long, as themselves to fall a prey." And the males? They, like elephants, leave their mothers and siblings in the female group once they reach adolescence. As they pass through adolescence and into sexual maturity, they gather into more temporary and loosely affiliated bachelor groups. Ishmael tells us, accurately I believe, that if you attack a "Forty-barrel-bull—poor devil! all his comrades quit him." (However, males have been known to strand themselves on beaches and die for no apparent reason other than to accompany their stranded and dying comrades.)[29]

As the males approach thirty years in age, they are large and confident enough to move down to the breeding grounds of warmer latitudes. There they swim by themselves, roving among the female groups while producing a distinctive and powerful *slow-click* vocalization: a call to fertile adult females, probably, and a warning to other adult males. They're looking for mating opportunities, and one will sometimes battle another male in a major physical contest over such an opportunity.

Humans have a distinctly more moderate sexual dimorphism than either elephants or sperm whales, with human males being on average only about four and a half inches taller than females, with somewhat denser bones and a higher ratio of muscle to nonmuscle tissue.[30] That moderate dimorphism suggests physical competition among males has been less significant in our own evolutionary history, perhaps supplemented by competition in other domains. But what other domains are possible?

One is sperm production, such as that characterizing the competition of some large and prehensile-tailed South American monkeys called muriquis. Muriquis are peaceful and seemingly egalitarian, and your first naive impression could be that somehow evolution has just left these beautiful monkeys alone. When I saw them in a Brazilian forest about twenty-five years ago, only a few hundred were still alive in the world, so, sadly, they are on the brink of nonexistence. In any event, scientists report watching a group of males peacefully taking turns to mate with a fertile female in an extraordinarily quiet and orderly session where no individual male seemed to rush or try to crowd out any other male. No one pushing to be first. No fights. No overt competition. That level of male calmness in the vicinity of sexual opportunity would never happen among elephants or chimpanzees, and you might fairly ask how anything about this calm, turn-taking sexual behavior could be considered competitive. In fact, the real competition happens at the level of sperm production and is identified by biologists with the term *sperm competition*. And just as physical competition tends to favor the evolution of large and aggressive males among elephants, so sperm competition among muriquis has produced males who have enormously large testicles. In addition to high levels of sperm production, muriquis have evolved a sperm that clots, so that once a male has mated and left a deposit of sperm that could potentially fertilize a female, it forms a clotted plug in the vagina that will tend to block the sperm of the next male. Muriqui males understand this problem at some level, however, so before one copulates, he is likely to reach into the female's vagina with his hand and simply remove the plug left by his immediate predecessor.[31]

Males compete against other males following the quantitative strategy for reproductive success in other domains also. There is the political domain, for example, which Frans de Waal introduced compellingly in *Chimpanzee Politics* (1982). Here we find the provocative vision of scheming male apes who achieve dominance and reproductive success by creating useful friendships and important alliances. And just as physical competition will leave its mark in big bodies, and sperm competition in big testicles, so political competition could be an important factor in the evolution of big brains—brains large and complex enough to enable their possessors to read the intentions of others, to outsmart and outmanipulate them with Machiavellian intent.

And the females? The qualitative strategy that generally marks female-to-female competition can emerge in a couple of ways. For one thing, females can try to mate with the higher-quality males and thereby increase their own probability of producing higher-quality offspring. Among elephants, researchers believe, the females actually prefer to mate with the biggest musth bulls. When one of these enormous creatures arrives on the scene, some females seem to act solicitous, even flirtatious—for instance, bumping against him or urinating directly in his path—and the one he actively seeks, the estrous female, is calmer than usual, less likely to bolt in his presence. Maybe his size merely intimidates these females, as it does most of the males, or maybe the females actually prefer a big male reeking of the musth scents that he carries along and even actively wafts into the air by waving his ears. Female elephants do have some final choice in mates, since having sex requires a male to raise himself precariously onto his rear legs, balance himself delicately on the female's back with his front legs, and astutely find her abdominally located vaginal opening with his own long and S-shaped erection. She can interrupt that fragile procedure simply by moving at an inopportune moment.

We might say that nature produced those giant males. The filtering mechanism of evolution preferentially selected bigger males over smaller males: the process Darwin called *natural selection*. But we can recognize that female elephants, by deliberately choosing the bigger males over the smaller ones, are also themselves promoting bigness among males. That process, which we might think of as a specialized subset of natural selection, Darwin called *sexual selection*.

Of course, natural selection has also affected the females, promoting any physical and temperamental aspects that are likely to follow the qualitative strategy by making them better at mothering. Among the most important temperamental aspects are those associated with the attachment virtue of empathy. A young mammal's life depends on his mother's responsiveness. As a result of this critical dependency, evolution promoted, in mammals especially, a supremely responsive attachment between mother and infant, an automated sensitivity of one to the moods and needs of the other. Indeed, empathy is closely associated enough with being female in our own species that one good way to distinguish newborn boys from newborn girls is to notice who cries more often in response to the crying of other babies. Empathetic contagion is stronger and more predictable for females, even at birth.[32]

I can recall riding on the back of a large female elephant through the precarious switchback trails of mountainous western Burma, on my way to watch the timber elephants at work, when, quite suddenly, everything stopped. The three adult female elephants of our expedition just stopped moving, and those of us sitting on top, including the mahouts, could do nothing about it, could only lift our heads and enjoy the cooling breeze and leafy scenery. What had brought this entire column of massive, moving, swaying flesh to such an abrupt stop? I looked down to see. One of the baby elephants, toddling along in the same group, had decided he was hungry. He wanted to nurse, so we all had to wait.

Describing a stereotype about gender differences among elephants that reasonably summarizes the known facts, I might say that the females value social and family attachments more than the males do. The males, by contrast, appear to be less social, more likely to be solitary. And in their ordinary bachelor lives they relate to one another in a distinctly hierarchical fashion: a system that ranks authority and sustains the rules that accompany it. According to Caitlin O'Connell-Rodwell, who has studied savanna elephants in Namibia for the last seventeen years, the males "display fascinating ritualistic behaviors" while communicating rank. The younger males defer to the power of a big bull by lining up and, one at a time, reaching up with the tips of their trunks to the giant male's mouth. Trunk tips are supremely important and vulnerable. Mouths can taste and bite. It's an "invitation to interact," according to O'Connell-Rodwell, as if the less dominant males are saying, "I see you. I am submissive, and I won't cause any trouble."[33]

Ordinarily this system works well for elephants. There is a division of labor, a duality of attention; and the matriarchal family remains a powerfully nurturing center. When the family group is threatened by potential danger, all the adult females will bunch together and form a heaving, collective fortress of flesh and trunks and outward-pointing tusks. This bunching behavior, which places the powerful adults at the outside and the vulnerable young at the center, is ordinarily an effective defense against predators. It can have the opposite effect, though, when the predators are modern humans with modern guns.

One notable approach was originally perfected in Uganda by Ian Parker, the founder of Wild Life Services. The services he offered were killing ones—called *culling* to emphasize their legal and supposedly rational nature. With a small group of hunters arriving on foot, Parker would carefully move close to an elephant family and then, at the right moment, startle the group with a sharp sound: a cough, perhaps, or the sound of metal striking metal. The elephants, instantly alert and soon alarmed, would gather into the defensive bunch, whereupon the hunters would spread out in a semicircle and, on signal, fire their automatic weapons. The matriarch and any other big adults were shot first, and once they fell, the rest of the animals, suddenly exposed and confused, could quickly be brought down. It might take only half a minute to finish off an entire family.[34]

Culling was a way of keeping the elephant numbers under control so that all the tourists coming to the national parks could see the right number of elephants rather than too many. Too many elephants, like too many people, will also have a disruptive effect on the landscape.

In Zimbabwe, Adrian Read applied some of Parker's techniques for his own government-sponsored culling operation. Read's team would consist of three hunters, each man carrying a semiautomatic rifle and backed by a gunbearer with a second loaded weapon. Overhead, a plane chased the elephants into the right spot, where they then bunched defensively—and behind the hunters another 250 men would follow in trucks for the efficient processing of dead elephants into meat, leather, ivory, and other marketable products. Culling in Zimbabwe, then, controlled elephant populations, provided employment for the rural poor, and earned money that was supposed to help the national parks. Elephants were paying for their own upkeep and improving the economy of struggling Zimbabwe simultaneously, as, similarily, they have theoretically been doing more recently in several southern African nations, including South Africa.[35]

The controlled and legal culls in southern Africa, combined with an uncontrolled and criminal continentwide killing of African elephants for meat and ivory have devastated elephant numbers globally: half today what they were a generation ago, one twentieth what they were a century ago. That's what I call *the elephant holocaust*. It is a cataclysmic slaughter that is at once extinguishing a magnificent species while simultaneously impressing some strange and terrible effects on the minds of the bereft survivors.

The psychological effects scientists have only lately and tentatively begun to recognize, although the sense of something gone wrong among elephants has been developing for at least the last couple of decades. Since the beginning of the 1990s in Pilanesberg National Park, the Hluhluwe-Umfolozi Game Reserve, and a number of other places in South Africa, for example, young elephant males have been raping and killing rhinoceroses. The rapes are bizarre, while the killings are just abnormal, as is the increased number of killings by elephants of other elephants elsewhere in the region.[36] In South Africa's Addo Elephant National Park, as many as 90 percent of recent elephant deaths have been caused by other elephants—a shockingly high rate compared to the baseline of around 6 percent for any normal elephant community. In the summer of 2005, meanwhile, wildlife managers of Pilanesberg finally shot three young males who, after having killed sixty-three rhinoceroses, had turned their aggression against tourist safari vehicles.[37]

Elsewhere across elephant territory in Africa and southern Asia, we find increasingly distressing reports of elephants gone mad (a psychological assessment), or bad (a moral one), and attacking not only elephants but also humans, devastating crops, killing farmers and hunters and tourists. Yes, it could be that the distressing stories of human-elephant conflict are the sad and simple consequence of humans encroaching on what was recently elephant territory. Humans living in those parts of the world that include elephant habitat have doubled in number during the last twenty-five years, a trend that neatly coincides with a halving of the world elephant population during the same period. But "Elephant Breakdown," a 2005 essay published in the prestigious scientific journal *Nature*, considers a deeper possibility: that the elephant holocaust has produced a new generation of animals traumatized by the violence they've witnessed and additionally disturbed, psychologically and perhaps morally, by the resulting collapse in normal elephant family structure.

The essay was collaboratively written by psychologists Gay Bradshaw and Allan Schore with three elephant experts—Janine Brown, Joyce Poole, and Cynthia Moss.[38] Gay Bradshaw would later tell Charles Siebert, a writer for the *New York Times*, that she had originally been moved to explore the issue because she saw a disturbing "intentionality" manifested in the recent reports of elephant violence. "What we are seeing today is extraordinary," she went on. "Where for centuries humans and elephants lived in relative peaceful coexistence, there is now hostility and violence."[39]

Psychologist Schore, an expert on early brain development and the neuroscience of trauma, additionally noted the importance of early relationships with the mother. "We know that these mechanisms cut across species," he told Siebert, adding that the brains of both humans and elephants are, during the early years, profoundly affected by certain "attachment mechanisms." The normal experience of attachment or bonding with the mother leads in turn to the development of an emotional brain with a greater capacity for attachment, that is, "greater resilience in things like affect regulation, stress regulation, social communication and empathy." But when the early experiences of maternal attachment are traumatically interrupted, for the developing young "there is a literal thinning down of the essential circuits in the brain, especially in the emotion-processing areas."[40]

So early bonding experiences with the mother are critical for the development, for both humans and elephants, of a neurologically derived attachments capacity: the basis for what I would call *attachments morality.* But the *Nature* essay also attributes a great importance for "socialization" and the role of fathers—or, since elephants most probably never know who their real fathers are, the older males. This may especially be the case for younger males, whose brains develop more slowly than their female counterparts' (a pattern typical for mammals generally). The socialization experience for both males and females, and the complex development of what I call *rules morality,* begins early on in the family dominated by mothers, grandmothers, aunts, older sisters, and female friends. When the males leave this social group at adolescence, though, their brains are also moving into a second phase of development that is affected by their new social experience in the all-male bachelor herds.

We are left wondering what kind of social learning and neurological development takes place during this postadolescent period, in normal elephant society. But I would like to suggest that this, too, may be an important period for rules learning and enforcement, particularly of the rules having to do with authority, violence, and sexuality. Obviously, that idea is complete speculation, hard to prove or disprove, but I find it interesting that when park managers introduced older males into those South African parks where the younger males had been so violently destructive, the effect was dramatic on both behavioral and biochemical levels. The presence of those older bulls ended the young bulls' hyperaggressiveness, and it also terminated their abnormally early and extended musth periods of high testosterone.[41]

CHAPTER 13
Flexibility

Starbuck (to Ahab): "Oh, my Captain! . . . let us fly these deadly waters! let us home! Wife and child, too, are Starbuck's . . . even as thine, sir, are the wife and child of thy loving, longing, paternal old age! Away! let us away!—this instant let me alter the course!"

Ahab (to himself): "What cozening, hidden lord and master, and cruel, remorseless emperor commands me; that against all natural lovings and longings, I so keep pushing, and crowding, and jamming myself on all the time; recklessly making me ready to do what in my own proper, natural heart I durst not so much as dare? Is Ahab, Ahab? Is it I, God, or who lifts this arm?"

—Herman Melville, *Moby-Dick*[1]

Is Ahab, Ahab? The captain indulges himself, here, in an entirely uncharacteristic act of introspection.

Ahab is not a navel-gazer, and for him to ask questions about the nature of his own mind and motivation might be considered a lapse, a case of serious if passing weakness. We can say, in any event, that the captain is somehow compelled—late in the voyage—to pause and wonder. What is the origin of his own often painfully felt behavior? Why does this mad killer's chase in which he has been so single-mindedly engaged feel so much like a violation of the most settled and steady impulses of his deepest self, those "lovings and longings" that appear to reside in his "own proper, natural heart"?

The image of the divided self is a modern one. You and I, living in a modern age, are used to thinking of people as psychologically complex, as constructed in ways that inevitably produce a steady parade of conflicting inhibitions and inclinations; and so for us Ahab's interior conflict is unsurprising. But the division he describes is still distinctive, and it's thematically introduced at the chapter's start by a strange vision of nature itself divided.

The chapter opens on an unusually beautiful morning, with a serene, bright blue sky, a calm breeze—and a world starkly split between the masculine and the feminine. The twin hemispheres of sky and sea are merged with a single compelling color, an "all-pervading azure"—but the "pensive air" is "transparently pure and soft with a woman's look," while the "robust and man-like sea" rises and falls "with long, strong, lingering swells, as Samson's chest in his sleep." The language is poetically sentimental, but the meaning is prosaically clear—and just in case we missed it, we are returned a second time in the second paragraph to this vision of a divided world: where small, "snow-white" birds glide on the air above like "the gentle thoughts of the feminine air," while far below them, in the fathomless sea, turn the great and destructive creatures of the sea—whales, swordfish, sharks—which represent "the strong, troubled, murderous thinkings of the masculine sea."

Ahab, meanwhile, has gone soft, and we soon see him longing for his beloved wife and child, wallowing in the emotions of attachment with home and family. Leaning over the side of the ship, he actually drops one anguished tear into the ocean.

Enter Starbuck.

When Starbuck quietly appears on deck and approaches the captain, Ahab is moved to confide. He confesses his regret for the life he has chosen, his anguish at missing wife and child. Wife? More like widow. He left but a single indentation in their marriage pillow, Ahab declares poetically, before shipping out the next day. It's a description of longing that Starbuck can readily identify with, since he, too, has left a beloved wife and child back home—but Starbuck also appears to recognize an opportunity.

From the start of this voyage Ahab has shown himself to be unfit for command, a frightening man whose mentally unbalanced state endangers the life of everyone aboard the *Pequod*. Now, with the captain temporarily weakened and hesitating, momentarily questioning his own purpose, Starbuck astutely moves to encourage and build on this new line of thought. He plays upon Ahab's sense of loss, his longing for the sympathetic attachments associated with home, with wife and child. "Oh, my Captain!" Starbuck pleads, "let us fly these deadly waters! let us home!" He reminds Ahab of their shared interests: "Wife and child, too, are Starbuck's . . . even as thine, sir, are the wife and child of thy loving, longing, paternal old age!" Starbuck repeats his impassioned appeal: "Away! let us away!—this instant let me alter the course!"

This is the second and last time the first mate addresses his captain in such a frank and fervent manner, pleading with him to change course. It's a moral argument, actually, the second of two between the staid and stable first mate and his dangerous captain.

The first moral argument I referred to in chapter 1. That's the moment when Starbuck blurts out his accusation that Ahab's pursuit of the white whale is an act of madness and blasphemy. I would now like to suggest that we can think of that earlier argument as a rules argument. Certain rules must hold true when you're acting against an inanimate object without mind or feelings, a mere crop to be harvested, which is how conventional Starbuck thinks of whales. Yet other rules will hold true when you're acting against an intelligent creature who resides in the same mental and emotional and moral universe as you do, which is Ahab's odd perspective. That's the debate, but since these two men have such radically different underlying assumptions about the nature of whales, this rules debate cannot easily be resolved—or rather it can be resolved only by a simple decision made by the man in charge, the tyrannical captain, who still holds the legal power to command the ship and throw anyone guilty of insubordination into the brig.

The second moral argument is an attachments argument. This is where Starbuck plays on Ahab's attachment emotions, pleads with the captain to change course and sail back to the place where love and empathy compellingly abide, to home, to the arms of his young wife and the needs of his precious child—and, incidentally, to feel empathetic pity for Starbuck, who also desperately longs to return to his wife and child. This second argument includes some possibility for satisfactory resolution because the two men might at least agree on the nature of their attachments, and attachments can sometimes override rules.

The promise of any moral argument—whether it's a fictional one between a first mate and his captain on a nineteenth-century whaling ship or a real one between two people, each hoping to persuade the other about some pressing matter in the twenty-first century—is that the argument can work. That someone's mind can be changed. That moral flexibility is genuinely possible. Since arguments are based on complex symbolic language of the sort that only humans have, it's fair to expect that only humans can hope to change each other's mind in this fashion. But are humans actually able to change minds about any moral issue?

I say they are, even on a grand scale, and to support my case, I will point to historical evidence. The biggest moral argument of Melville's era was over slavery. Was holding slaves a moral thing to do? Or was slavery simply and clearly immoral? You and I inhabit a world far enough distant from the time when slavery was commonly accepted that we can see only a single moral possibility. Even to whisper the thought that slavery might in some circumstances be "right" or "good" is to invite passionate moral outrage from everyone around you. In other words, that moral argument is over, so completely finished that you and I have trouble imagining there ever was one. So we look back on the slaveholders of the pre–Civil War American South with scorn, disbelief, even hatred. How could any moral person do such an utterly immoral thing? And when we are reminded that even Thomas Jefferson, a principal author of the Declaration of Independence and third president of the United States, held slaves, we're puzzled or even shocked.

Jefferson appears to have been a pragmatic gradualist, at once recognizing the evils of slavery while also realizing how fully entrenched the institution had become in America, and thus believing, finally, that to end it would require time and, possibly, war.[2] But let the historians explore any finer details of Jefferson's thinking on the subject. The simpler point I want to make is this: Well into the middle of the nineteenth century, respectable men held opposed or nuanced views on slavery and debated each other on the subject. A century later, such was not possible. What happened? How did this moral argument get resolved?

The slavery argument took place in late-eighteenth- and early-nineteenth-century Euro-American society, when most people believed that all morality comes from a Christian God speaking through the text of the Bible. At first glance, unfortunately, that God appears to condone slavery, and anyone making a rules argument in favor of slavery would have several reference points in that Holy Text with which to buttress the argument and pound the pulpit.

I'm thinking here particularly of an American theologian and minister from Huntsville, Alabama, the Reverend Fred A. Ross, who routinely cited God's rules in support of slavery. God, Ross told the General Assembly of the State of New York in an 1856 address, demands obedience above all, and Adam's willful disobedience is the original sin, the source of all other sins. Adam ate of the forbidden fruit, and thus God condemned him and his progeny to endure pain and death in a fallen world where social relations would forever be characterized by "the law of the control of the superior over the inferior."[3] This summary rule, a simple variant of the obedience rule of the Ten Commandments, is manifested both by the institution of marriage, in which women are commanded by God to obey their husbands without question or complaint, and by the institution of slavery. God's support for slavery, Ross continued, is unmistakably spelled out in numerous passages in the Bible, particularly those found in the books of Exodus, Leviticus, Colossians, and 1 Timothy.

The Hebrews of Moses' time had fled from slavery out of Egypt, but they also kept slaves, so among the instructions communicated from God to Moses to the newly gathered tribes of Israel were those describing the proper treatment of slaves. The Bible is casually clear on this subject. For example, we can read in the twenty-fifth book of Leviticus, "As for your male and female slaves whom you may have: Your male and female slaves are to come from the nations round about you. You may also buy from among the strangers who sojourn with you and their families that are with you, who have been born in your land; and they may be your property. You may bequeath them to your sons after you, to inherit as a possession for ever; you may make slaves of them." A common tenet of Christian theology is that God's original harshness was miraculously softened by the appearance of a son and the empathetic New Testament teachings he inspired. But as the Apostle Paul advises in the New Testament (Colossians 3:18–22), the overarching rules regarding obedience to proper authority are still perfectly clear and inflexibly strict. Just as everyone must obey God, so, too—Paul writes—wives must obey their husbands, children their parents, and slaves their masters, and they all must do so "in singleness of heart, fearing the Lord."

But if the antislavery proponents found their arguments weakened by biblical rules that seemed to justify slavery, they had attachments morality on their side. I believe that any full historical analysis of the contents of the antislavery arguments and propaganda during this period will find that the most consistent theme was not a reference to moral rules, biblical or otherwise, but rather an appeal to moral attachments. More particularly, the antislavery argument worked largely by appealing to empathy, to a person's feelings of compassion in response to the genuinely horrific suffering endured by African slaves during the Middle Passage—crossing the Atlantic chained in the holds of slave boats to America—and then, in the American South, at the hands of slaveholders and their often brutal enforcers. In America, the attachments argument against slavery was by midcentury brilliantly summarized and dramatized by Harriet Beecher Stowe in *Uncle Tom's Cabin*—published in 1852, the year after *Moby-Dick* appeared. Stowe's sensational, bestselling novel, and the play based on it, moved millions of Americans to experience a passionately empathetic response to the sufferings of slaves, and in that way it helped convince them that slavery was a moral wrong of the greatest seriousness and urgency.

The ability to talk and write and otherwise communicate symbolically gives our own species an unusual potential for moral flexibility. As symbol-using creatures, we can discover new information, assess it, and realize new ways of considering things, and in that fashion the moral conversation continues. Such is the promise of history: that we can progress through knowledge and cultural agreement into a more perfect understanding of ourselves and a more finely tuned sense of our own morality. The power of language alone can explain the unusual flexibility of human morality . . . but language is not the only route through which moral flexibility appears.

Flexibility of a sort also emerges through the normal workings of variety. In human societies, we see a significant cultural variety. True, the underlying emotions responsible for moral systems are the same across cultures, but we produce cultural variants of those systems—so that, for example, some societies are more authoritarian than others, more severe in their enforcement of the rules, less tolerant of conflicting attachments, and so on. Such is the power of culture and tradition. And since rules are established and supported not by all individuals equally, but rather by a set of socially powerful individuals, we find a secondary source of variety routinely produced by the varying character of our leaders. It does matter, after all, whether your society is led by Winston Churchill or Adolf Hitler—just as, for one well-studied chimpanzee community enduring in a tiny forest of western Tanzania over the last several decades, it mattered whether the alpha male was Mike the clever consensus-builder or Frodo the hulking bully.

"It is amazing what a difference the personality of the alpha male can make in the relationship dynamics of an entire group," writes primatologist Susan Perry in *Manipulative Monkeys* (2008), describing the white-faced capuchins she has studied for fifteen years in the forests of Lomas Barbudal, Costa Rica.[4]

Perry recalls, for example, that scar-faced, ever-snarling old fellow she named Curmudgeon. As alpha male, Curmudgeon was "the obvious favorite" of all the females, who often groomed him; and wherever Curmudgeon chose to wander, he would be followed by a procession of "excited fans." Most of the fans were infants or youngsters, who made a friendly gargling noise as they approached and reached out, trying "to touch their idol"—while Curmudgeon contributed to the effect by his hair-raised swagger and lordly airs. He was "unquestionably the most self-confident alpha male I had ever known," Perry writes, and he always seemed to have a powerful sense of his own drama. Whenever he urinated, for example, Curmudgeon "arched his back, fluffed up his hair, made a resonant rhythmic grunting sound, and splashed urine all over his hands and the monkeys in the vicinity. None of the other monkeys looked anywhere near as important when they urinated."[5]

Curmudgeon's leadership style was typical for alpha male capuchins: expressively dramatic and obsessively tyrannical. Most of them, Perry writes, are "control freaks, constantly monitoring, commenting on, and manipulating the relationships of the subordinate males in their group."[6] Yet Pablo, who was the alpha of another social group in the same forest, appeared to have a remarkably different personality. Pablo was an easygoing, laid-back kind of guy who usually seemed to be unfazed by friendly interactions among his subordinates. In Pablo's community, therefore, males frequently groomed one another, while he himself was an unusually active groomer, grooming not merely females but also his male allies and even many of the juveniles. Perhaps most remarkably, Pablo would allow sexual contact between the subordinate males of his community. Male homosexuality was definitely a no-no in the other social groups of Lomas Barbudal.[7]

Whenever she switched from watching Curmudgeon's group to Pablo's, Perry felt she was "entering a different culture." Because Pablo was so relaxed with his male subordinates, they seemed comfortable in their mutual associations, and that in turn meant they were "highly cooperative" and "able to form a united front against enemy males that is frighteningly effective." Meanwhile, and possibly because they were so tightly bonded, the males in Pablo's group would occasionally coerce females sexually, which was virtually never seen in the other capuchin groups.[8]

These bare-faced monkeys have the largest brain-to-body ratio of any primate other than humans. They are sophisticated tool-users and omnivores who prey on other animals to get meat, which they share with one another. They also cooperate in hunting and in caring for their infants; and like chimpanzees, they cooperate to make war on others of their own kind—that is, other white-faced capuchins living in other social groups elsewhere in their big, dark, leafy world. So you would be wrong to imagine that they carry out their lives like little furry machines, without awareness or calculation or memory, without variety, choice, or behavioral flexibility. They are inventive and creative animals, highly manipulative—both of their environment and each other. They live in a social world that in some ways seems profoundly odd and alien to us, but it still includes many aspects that feel perfectly familiar. It's a world where friends and enemies abound, and where alliances form and shift, and one's social status is always precarious. Well, *precarious* is true for the males, anyhow. The females form stable dominance hierarchies. The males scheme and fight and engage in politics to determine their social status, with the alpha climbing to the top of the male hierarchy via ambition, nerve, and savvy.

Since the temperament of leaders can vary—a Hitler or a Churchill, an Ahab or a Starbuck, a Curmudgeon or a Pablo—then, to some degree, we can seek in that variety some potential for flexibility in the nature of society. Change that strutting character in charge of things, and you might in part change the character of a society. That's the eternal hope of the politically engaged.

Yet changing the individual in charge, whether through election or coup d'état or some other means, is not so easy. Deliberately or not, powerful individuals rely on the psychologically embedded tendency in others to submit to their authority. That impulse to submit is so deeply entrenched in our own emotional makeup that we seldom question it. Submit to the authority of the priest? The president? The king? Of course! Why would you not? But those who have gained power also rely upon social enforcements to hold on to it. Social enforcements: displays of strength, acts of intimidation, distributions of favors, strengthening of alliances, expressions of loyalty, laws prohibiting treason or mutiny or sacrilege, and an army or police force sworn to protect one from coups d'état.

Animals don't have written laws prohibiting treason, obviously, but they may well attend to important alliances or promote significant expressions of loyalty. Among the white-faced capuchins, males do a number of strange things that could reinforce friendships and political alliances. They sometimes suck lingeringly on each other's ears, tail, and fingers, for example. But possibly the oddest ritual of all is eyeball poking, where one monkey will draw another's hand up to his or her own eye, then guide the other's fingernail and then finger gently into the eye socket behind the eyeball. Mutual eyeball-poking sessions among these monkeys can last up to an hour, and they mean—what? Perry believes it could be a way of communicating a deep level of mutual trust.[9] Eyeballs are supremely vulnerable to damage and infection, after all, so it's a bold act demonstrating great trust to encourage another individual to reach a fingernail and finger into the highly sensitive area behind it.

Olive baboon males in Africa do something roughly comparable when coalition partners greet one another by reaching out and grasping the other's testicles. The males of this species are not ordinarily touchy-feely sorts; they never groom each other, for instance. So this mutual testicle-handling, with one bloke voluntarily placing his entire reproductive future in the hands of another, seems to express most emphatically a mutual trust.[10] One can imagine, if these baboons had language, the compelling poetry of their oaths: *By my very balls, I pledge. . . .*

Yet loyalty has limits, and oaths will be broken. Adolf Hitler survived seventeen serious assassination attempts. Captain Ahab survived disloyal whisperings and mutinous plottings aboard the *Pequod.* No leader can survive forever, though, and without a formalized, written method for changing authority, the powerful are often doomed to endure the informal and unwritten methods for social change, such as assassination, mutiny, or coup d'état—as Curmudgeon abruptly discovered one day.

After a brief trip to San José, the capital of Costa Rica, to renew her research permit, Susan Perry returned to the Lomas Barbudal forest to find Curmudgeon alone, injured, and emitting only a pathetic series of *I-am-lost* calls. His voice, Perry writes, "cracked and trembled, each syllable ending with a whiny whimper, in stark contrast to the confident, steady quality his voice had had before." She had never before seen him alone. She had only once seen him injured. But now he had a deep gash on the bottom of one foot, another on one shoulder, and puncture wounds elsewhere. Unable to use his legs, the former alpha moved by dragging himself with his arms up trees and across branches. His hair was flattened, making him appear "deflated" in comparison to his previous hair-raised confidence, and altogether he just looked "miserable and terrified."[11] Curmudgeon had been displaced by a new alpha male, who must, Perry believed, have been supported by his own group of plotting allies. No single monkey, working alone, could have done such damage. Poor Curmudgeon!

The problem with coups d'état is that they change individuals without changing the nature of their power. In a typical coup d'état you merely exchange one guy, or a cabal of guys, for another. Sure, superficial differences will exist between this leader and the next one. The differences might even be important. But ultimately, if you merely exchange one leader for another, you can expect, on average, the same temperament at work within the same moral system. Without structural alterations, there can be little hope for true flexibility or enduring change. To discover real change in any human or animal society, one looks not for the common, ordinary coup d'état, but rather for the rare and extraordinary revolution. We might define *revolution* as an event that significantly alters the way power is distributed in a social group.

We are fortunate to have a good animal model of a real revolution: the ancestral split and subsequent developments that led to modern chimpanzees and modern bonobos. DNA analysis tells us the following. The ancestors of chimpanzees and bonobos were once members of a single population of forest-dwelling apes. Then, around 2.5 million years ago, that ancestral population was divided geographically. The two groups stopped seeing each other, stopped mating, stopped mixing genes. They evolved separately for long enough that they became two separate species, living today on separate sides of the great Congo River that cuts an enormous curving arch across the forested center of the African continent.

The revolution that happened was, in part, sexual, with a remarkable shift in the power that one sex often maintains over the other, and we can see the results today. These two closely related species, chimpanzees and bonobos, are today patriarchal and matriarchal respectively. Considering the social systems of numerous other species that branched away from the same ancestral tree in earlier times—all the other apes, nearly all the other primates—we can confidently presume that the common ancestor of chimps and bonobos was a forest ape living in a patriarchal society. The matriarchy of bonobos is the interesting exception, and we should look to bonobo history to consider the revolution that must have taken place.

To describe any society as patriarchal or matriarchal is merely to assert that one sex predictably dominates the other. More abstractly, we might imagine that one sex or the other is in charge of the rules and rule enforcements. The question then becomes, which sex? When valuable food is openly available, which sex gets the first and best pieces? When a male and a female of the same age moving down a narrow trail come face-to-face, which one predictably gets out of the way? In any contest over any resource—be it physical or territorial or social or reproductive—which sex predictably gets the upper hand?

Since Japanese primatologists have maintained some of the longest-term research sites for both chimpanzees and bonobos and, at both sites, have used the same technique of baiting their observation areas with fresh sugarcane to attract the wild apes, perhaps the Japanese results give the most direct comparison of these two species on matters of dominance expressed in access to food.

At the Mahale Mountains National Park research site in Tanzania, where the Japanese have watched chimpanzees continuously since the mid-1960s, researchers note that the chimp males routinely dominate females during feeding. If a female enters the feeding area and notices some extra sugarcane treat that is partially hidden by brush or vegetation, for example, she may simply move past the area, as if pretending that she has seen nothing at all interesting. Then, once any nearby males have wandered off, she will return and feed on the sugarcane. "Nevertheless," as primatologist Takayoshi Kano summarizes, "an ill-tempered male who has been keeping an eye on the female's behavior may fiercely attack her and steal the sugarcane she was holding."

At the Wamba research site in the Democratic Republic of Congo (DRC), where a Japanese team under Kano's direction watched bonobos for two decades, from the mid-1970s to the mid-1990s, dominance patterns at the feeding area looked absolutely different. A moving group of bonobos would enter the clearing, with the lower and middle-ranking males arriving first. Those males would eat what they could. Soon after their arrival, though, the rest of the group would emerge from the forest, including the high-ranking females and some high-ranking males, whereupon the first males grabbed any pieces of sugarcane they could and scattered away. The higher-ranking males would remain at the site, eating sugarcane, but—unlike with chimpanzees—the females would act entirely unconcerned by their presence. They would amble right up to the males, pick up any food in the vicinity, and either sit down and eat right there or carry off their own morsels of sugarcane to a more convenient spot. "When they have plenty of time for sugarcane," Kano writes, "males and females appear very much at ease with each other."[12]

But what if the sugarcane is in short supply? Then, even the high-ranking males seem wary at the approach of the females. These socially important males might stand up, use their hands to hold on to whatever pieces of the food they can, and then, turning and walking upright, disappear back into the forest.

European and American scientists working at Lomako, a second bonobo research site in the DRC, documented the way these apes divide up rich food prizes: the occasional piece of meat or some large fruits, such as the wild jackfruit, *Treculia africana*. The appearance of such a treat causes bonobos to react rather as chimpanzees do in the same circumstance: someone possesses and claims ownership of it, a crowd of begging supplicants gathers, and the owner eats while doling out a few pieces to some of the supplicants. With chimpanzees, the lucky owner is almost always a male, while with bonobos, so the Lomako data show, the owner of the prize is almost always female.[13]

We can conclude from these and a number of other studies and observations that bonobos are indeed a matriarchal species. Through their generally higher authority, females define the rules and thereby control access to important resources, such as food. But how is this possible?

It's certainly rare among mammals, where only a small handful of species are, like bonobos, decidedly matriarchal. The reason matriarchy is so rare is not hard to discern. Social power is often sustained by physical power, and the males of almost all mammal species are larger and more powerfully muscled than their female counterparts, and they often carry more fully developed weaponry, including horns, tusks, and canine teeth. That's the ordinary form of sexual dimorphism for mammals. On average, males are also temperamentally more violent than females. This temperamental difference we can see even among the peace-loving bonobos. A study of 325 acts of aggression among the Wamba bonobos found that fewer than 8 percent of the aggressors were female.[14] Yes, the females do fight on occasion, and when they fight each other, they do so quickly and often quite ferociously. One female may without warning leap on top of another, bite her, and take away her food, and then, sometimes, the two end up rolling around on the ground. But the males fight each other far more often, and their attacks are slower to develop. A conflict between males starts with the two glaring at each other, then circling and threatening and glaring some more, until finally they fight.[15]

That the bonobo males seem, like males in many other species, so much more inclined to settle things by violence—and they are, after all, bigger and stronger than the females—just adds to the puzzle of bonobo matriarchy. Among spotted hyenas, a matriarchal species outside the primate group, females have climbed to the top the old-fashioned way: They have evolved to be bigger than the males. The size difference of around 12 percent remains stable even in captive groups, where food is given equally to both sexes.[16] Beyond this size difference, females are also temperamentally more aggressive, even at an early age; their high levels of aggression are undoubtedly related to their unusually high levels of male hormones, such as testosterone. (Indeed, the effects of abnormally high androgen levels in pregnant mothers on fetus development probably helps account for another unusual fact about spotted hyenas: the females are sexually masculinized, which means their genitalia are virtually indistinguishable from that of the males, with clitorises that look identical to penises and labia resembling scrota.)[17] But hyena females have yet another advantage over the males, in that their social system avoids incest by having males migrate away from their birth clans, while the females remain. As a consequence, the adult females are naturally bonded as sisters and aunts, while the adult males, having emigrated from other clans, start as strangers and remain, to some degree, anxious outsiders with few or no blood relatives to count on. Here we can say that sisterhood rules.

Spotted hyena females thus have three kinds of advantage over their male counterparts: slightly greater physical size and strength, somewhat higher temperamental aggressiveness, and intrinsically stronger social bonds derived through genetic relatedness. Those are exactly the same three advantages that chimpanzee males rely upon to maintain their powerful political grip on their communities. But once again, bonobos seem to challenge this entire idea. Bonobo males have all three advantages that the chimp males and the spotted hyena females possess, yet it's not enough. What gives? Why aren't bonobo males in charge of things? What has happened in this species to produce such a radical transformation, a genuine revolution in the arrangement of social power?

Bonobo females have gained power through an acquired sisterhood. Although the adult females in a bonobo community start as outsiders, immigrating from neighboring communities and therefore not being genetically related to each other, they are still more strongly bonded than the males, and the female cooperation that comes from such strong bonds makes all the difference. Female cooperation? The Wamba researchers describe those comparatively rare incidents when a male attacks a female, sometimes violently. But unlike with chimpanzees, the attacking male bonobo almost invariably risks a massed retaliation by angry females.[18]

When I call theirs an *acquired* sisterhood, you might imagine it has easily or simply been acquired. That it was a random cultural invention or some historical fluke, a fashion that at one historical moment just caught on among those smart little apes. The story is not at all that simple. Female chimpanzees occasionally do the same thing as female bonobos. They will sometimes mob males who have become too aggressive. But they don't do so consistently. Why not? They are less often together and less tightly bonded. Among chimpanzees, the males are definitely the more gregarious sex, preferring to hang out together and showing the strongest social bonds, while the females are generally more solitary. Among bonobos, the situation is entirely reversed, so that the males are more solitary and the females more gregarious and bonded—with strong bonds of friendship that are strengthened by sex and what we might think of as romantic attachments. These strong bonds have, in turn, enabled bonobo females to counter male attempts to dominate through threats or violence.

There is power in numbers, but the fact that bonobo females have discovered that power still does not explain why they consistently use it. Probably the best answer is that they consistently use it because they can, and they can because their food supply allows them to. A steady food supply may have been the important ecological situation that enabled this major social revolution among bonobos.

Food supply can affect males and females differently. Like most mammals, chimp and bonobo females spend much of their adult life caring for their offspring, first with nursing, later with helping them find food for themselves. Whereas any adult, male or female, can survive with little food for a few lean days or even weeks, an infant or juvenile cannot, which means females for both species, to succeed reproductively, are still required to find a regular supply of food for themselves and their offspring. Both species rely on some highly nutritious but often spottily distributed sources of food—meat and fruit—but bonobos live in a part of Africa where they can almost always count on a third kind of food: some particularly nutritious herbs, such as the *Haumania liebrecht-siana*, that are distinctive for low amounts of indigestible fibers and high levels of protein.[19] Such nutritious herbs are common in bonobo forests, south of the Congo River, and uncommon in chimpanzee forests, on the other side of the river—and possibly this one critical difference in the forest vegetation is itself a result of gorilla distribution, since gorillas ordinarily feast on the same kind of ground-level herbs. But an extended warming trend that dried out the forests eliminated gorillas from the middle of

Africa at some time in the distant past, perhaps around 2.5 million years ago, so that even today there is a broad stretch of gorilla-less habitat in the middle of Africa. That large piece of gorilla-less forest is where the bonobos live, and it's where they are able to rely on a steady supply of nutritious gorilla foods.

This theory, promoted by anthropologist Richard Wrangham and detailed in our coauthored book, *Demonic Males: Apes and the Origins of Human Violence* (1996), points to the great warming trend and the gorilla die-off in Central Africa as the critical events that changed the ecology in these forests, producing the pattern of food supply that eventually separated chimpanzee-bonobo ancestors into two groups where a social revolution—females finding power in numbers—would occur in one of them.[20] Following and merging with that social revolution, a whole series of evolutionary changes would begin to take place among the bonobos, thus embedding the social event significantly in the species' physical and psychological nature. The result: a remarkable kind of ape, quick and bright and matriarchal, whose lifestyle can be characterized as one of significantly more sex and distinctly less violence.

Such is the complex interpenetration of ecological, evolutionary, social, and psychological events that characterizes all true revolutions. That we often think of human revolutions far more modestly, as strictly social events, could be an artifact of the usual historical time frame. History takes us back fifty years, a hundred years, two or three hundred years. But when we look at these events in a wider chronological context—maybe a few thousand years or tens of thousands of years—we can begin to see how ecological and evolutionary events merge with social and psychological ones. Let's imagine, for example, that the relationship between small bands of nomadic people and the rich soils and plentiful water of Mesopotamia led to settled agriculture and animal domestication. Those revolutionary changes in turn enabled humans to live for the first time in greatly expanded social groups, while greatly expanded group size may have been the driving force behind the first written moral rules, since informal enforcement based on unwritten rules may only be feasible in small communities where everyone can recognize everyone else.

The writing down of moral rules was another revolution, yet another important shift in the arrangement of social power; and the great forces set in motion by written codes may ultimately have influenced, complexly, much later events, such as the democratic revolutions of eighteenth-century France and North America. In these revolutions, an old-fashioned aristocracy—a small elite holding power and passing it from male to male through the paternal genetic line—was replaced by a new system of power based on written constitutions and a new machinery of popular participation: assemblies and votes.

The French and American revolutions were remarkable events, certainly, and they spread by imitation rapidly across cultures to move broadly beyond their origins in the Western world. They also (by establishing the machinery of participation through voting) began an even more radical and interesting sort of revolution. We might say this more radical revolution traces its origins back to Melville's time, more or less, but it remains incomplete. Global in nature, it now appears almost everywhere in the human social and political worlds, but it has taken root at differing speeds and with varying degrees of success. I'm speaking, of course, of the revolution that began—phase one—with the remarkable notion that women should have the right to vote. Among nations that still exist, New Zealand was the first, giving full voting rights to women by 1893. In Europe, the first was Finland, which gave women universal suffrage in 1905. Then it was Norway and Denmark in 1913 and, following the First World War, the Soviet Union, Canada, Germany, Poland, the United Kingdom, Holland, the United States, and Turkey. In succeeding years and elsewhere around the world, nation after nation followed suit until, by the start of the twenty-first century, only a tiny pocket of rigidily patriarchal communities—such as Brunei and Saudi Arabia and the Vatican—still withheld from women the right to vote.[21]

If the first phase of this revolution was giving women voting rights, the second phase will be the consequence of voting, which is political and thereby social power. The second phase of this revolution has hardly even begun to take shape, certainly, and so we can only roughly speculate about what it might finally mean.

A popular nineteenth-century notion held that if or when women got the vote, they would have a civilizing effect on men. That simplistic idea has a Victorian whiff to it, and in various versions it has reflexively been rejected by twentieth-century feminist theorists, who came to regard essentialist thinking about gender as regressive. Essentialist thinking: that inherent differences exist in the average psychological makeup of the two sexes. The usual argument against essentialism: that gender is really a social construction, a random invention of meaningless role-playing designed by patriarchal men to keep women weak and subservient. The idea of gender roles as nothing but arbitrary social constructions still has a good deal of popular currency, but it's wrong, and it leaves us with the following unsatisfactory thought. If women are, on average, temperamentally no different from men, what's the point (other than as a pleasant gesture of fairness) in giving them the vote? What was all that fuss about? The real hope, and the strongest reason to promote women's voting rights and, from there, increased political power, is based on a completely opposite notion about the meaning of gender. If, as I believe, women are on average temperamentally different from men, then one can hope that once they have real power, political and thus social power, they will eventually place their own positive stamp on the shape and nature of the human community at large. I will summarize this hypothesis as *the women's effect.*

What might this women's effect look like?

First, giving women power and social power should increase the force of the so-called *women's issues.* You and I could easily name a few such issues: predictable maternity and paternity leave from work, more opportunities for husbands to help with the parenting, more accessible child-care services, better health care for women, more effective ways to combat rape, greater individual choice in abortion, a woman's equal freedom to choose where to go and what to do, and so on. We might imagine that these and other women's issues have randomly been picked up by women's political groups for no particular reason. Alternatively, we might think that they are just obviously the sorts of things that females would be interested in. But why are they obvious? The evolutionary answer, which requires us to think of women as biological beings with minds shaped by an evolutionary past, is that these issues promote women's reproductive success through greater freedom from male dominance, increased choice of potential mates, and a qualitative investment in their offspring.

Put that way, women's issues may still sound for the most part entirely sane and obvious and ordinary. Only when placed in contrast to the notions and structures of highly patriarchal societies can we see how radical these issues are. I believe that patriarchal control characterized most earlier forms of human social existence, but to find fair samples of what they probably looked like, both historically and prehistorically, we needn't look far. We can simply consider those contemporary societies that for various cultural and historical reasons have been been remarkably slow to change. The obvious and defining mark of any highly patriarchal society is that men are in full charge of both the political and religious institutions.

Somewhat less obviously, these highly patriarchal societies tend to be strongly and rigidly hierarchical and rules-oriented, and they consistently favor male reproductive interests at the expense of female reproductive interests. In practice, this means controlling women's movements and appearance in public, how they dress, whom they may associate with and under what conditions, and so on. Such controls are often described by both men and women as being "protective," but their primary function is restrictive. They promote men's choice of mates at the expense of women's choice of mates. Most telling, however, is that traditional patriarchies often enable men at the top of the hierarchy to seek increased reproductive success through the quantitative strategy: multiple wives, which means the evolutionary winner's opportunity for unusually large numbers of offspring.

Now we can see why women getting the vote was such a struggle. Not a shot was fired, but this battle was real and ferociously fought. Why? Because the stakes were actually enormous. Many men were afraid that women would vote differently from men, and those fears were probably accurate. The women's revolution, when it is complete at some point in the future, ultimately means the end of patriarchy in human societies, and that will, in turn, produce a major shift in how we all listen to and understand our inner voices: our internal sense of what morality is and how it feels. In terms of rules morality, the shift might appear most clearly in the realm of sex and sexuality, which is where men and women often see the world most differently. Thus, we should begin to witness (in many places already have) a change in the general, or average, human notion about right and wrong in sexual behavior.

But along with a rules change, we can also expect a more embracing and amorphous change in the general expression of human morality, since the increased involvement of women may also bring a somewhat increased significance for attachments—particularly those associated with empathy—at the expense of rules. A human society in which men and women have equal power may be, on the whole, somewhat less dogmatic in judgment and somewhat more empathetic in action than those patriarchal human societies we already know about from contemporary examples and the historical record.

Oh, yes, this is all rampant speculation about medium-term social change, and I can't fault you for dismissing it as high-riding nonsense. You might say that the women's revolution has been much more pragmatic than all that. Or you might say that it's been a complete fizzle, that all the energetic fist-pumping and putative bra-burning of the 1960s has led to little more than high-heeled women entering high-level politics fifty years later while adamantly insisting that they are utterly "traditional": reassuringly old-fashioned mothers and old-fashioned wives. These particular assertions may or may not be true. You might or might not be right in the details. But to examine anything through a fifty-year lens is to see almost nothing. We are living near the start of this historical shift in the nature of human social power. We may not even be aware of it or recognize it as anything more than a temporary and possibly irritating little twist in the normal arrangements, and we certainly cannot be expected to see where these changes will take us over the next one hundred or one thousand or ten thousand years. Dreaming fitfully in our cloud of unknowing, we hardly recognize the past and barely understand the present. Our vision fails altogether as we try looking into the future.

CHAPTER 14
Peace

Retribution, swift vengeance, eternal malice were in his whole aspect, and spite of all that mortal man could do, the solid white buttress of his forehead smote the ship's starboard bow, till men and timbers reeled.

—Herman Melville, *Moby-Dick*[1]

Moby-Dick. You know how the book ends: cataclysmically. Whale strikes ship head-on (smashingly). Ship sinks (quickly). All the men drown (sadly), except (conveniently) for the eager young narrator, Ishmael, who lives to tell the tale. The End. The Crashing, Cataclysmic, Disastrous, and Pretty Damned Depressing End.

You might dismiss this whole story as a wildly fictional fantasy, an overly long piece of pretelevision entertainment with no basis in reality or truth. But in fact, *Moby-Dick* was based not merely on Melville's own life experience as a common seaman aboard a whaling ship; it was also inspired by a true story, the staving and sinking of the Yankee whaling ship *Essex* by a large sperm whale.

The sinking of the *Essex* was most famously described by one of the survivors, Owen Chase, in a 1821 book with a title almost long enough to serve as a high school book report: *Narrative of the Most Extraordinary and Distressing Shipwreck of the Whale-Ship Essex, of Nantucket; Which Was Attacked and Finally Destroyed by a Large Spermacetti-Whale in the Pacific Ocean.*[2] The *Essex* story was certainly known to Melville. He was also familiar with a contemporarily imagined sailors' yarn about another dangerous whale named Mocha Dick, an albino giant introduced to the American reading public as the eponymous hero of a piece of short fiction by J. N. Reynolds, published in 1839 in the *New York Knickerbocker.* Mocha Dick was a monstrous animal who habitually attacked his attackers—whalers, that is—and was "*white as wool!*"[3]

So, we might conclude, Melville's great tale is simply a hybrid mix, a lazy man's stew of refrigerator leftovers, an opportunistic pastiche of truth and fiction, or (to put it in a way guaranteed to please the English teacher) a marvelous construction of the imagination centered on a few provocative bits of truth, like sweet fruit around tiny seeds. Yet, the story is far truer than that if we can understand *Moby-Dick* more generally as an environmental manifesto, something like an enormous piece of stained glass, a glowing and intricate representation of the human war against animals and nature based on the following theme: *Bite nature, and nature will bite back.*

Our natural focus on Captain Ahab's dramatic craziness could easily mislead us into thinking that *Moby-Dick* is merely the tale of one strange and dangerous man who has been given command of a ship and wants to kill some imagined beast no matter what the cost. Someone rather like Adolf Hitler, the little man with the evil mind who captained for a few catastrophic years the German ship of state. Disturbing. Dangerous. Mutiny or assassination might be a good idea.

But to my mind the really dangerous character in this tale is Starbuck. Starbuck is far more worrisome than Ahab because he's entirely normal and perfectly conventional, an average man with an average perspective on the world. Starbuck thus most fully embodies the psychological nature of his species—or at least the male version of it. He's far more dangerous than Ahab in part because he represents the sort of man who would in reality have gained command of a whaling vessel, both in Melville's time and ours. Few people would plot a mutiny against Starbuck because most people would not think to question his style and motives. He's normal. He's professionally trained. He's passed all the tests and earned all the licenses. He is also convinced that he belongs to the only actually alive species on Earth, is prepared to kill (or rather, in his mind, "to harvest") every nonhuman creature. He sees himself as personally directed by the Divine Intelligence to exploit every living creature, every plant and animal, the water and the air, as routinely and efficiently and completely as possible. And since Starbuck most resembles the average professional whaler of his time and our own, we can credit Starbuck, or rather the real-life versions of him, with the historical progress of whaling as an industry.

Because they see whaling as a business just like any other business, such men have come to embrace every possible technical advance as simply one more profitable click in the ratchet of normal human progress. The arrival of steel ships: click. The adoption of steam, then diesel, engines: click. The invention in 1868 of the cannon-fired, exploding-grenade harpoon: click. The development, in the twentieth century, of giant factory ships with vast stern slipways and ten-ton pressure cookers: click.[4] The result has been progress, real progress in the business of killing whales. Americans of the nineteenth century dominated this industry, and with their square-rigged wooden ships, oar-powered chase boats, and hand-thrown harpoons, they killed some thirty-six thousand sperm whales in 120 years, an average of three hundred per year. That's what the industry was like in Melville's day. A hundred years after the publication of *Moby-Dick*, the Japanese and Soviet Union fleets alone were slaughtering whales at fifty times that rate.[5]

Such is where the Starbuckian approach to nature and animals takes us, and, yes, it has done good things for ever-needy humans. In the nineteenth century, it brought gainful employment and manly adventure to a few generations of working-class men, and it brought baleen corset stays and brightly burning candles to some middle-class women. In the twentieth century, it brought to yet other folks oils and lubricants and fertilizers, pet foods, people foods, and glycerin for the manufacture of explosives. Of course, the corsets long ago fell out of fashion, and the candles long ago burned up. The oils and lubricants have been ground down. The explosives have been exploded, the pet and people foods eaten. And now the remaining whales are going as well.

Nearly half of the thirteen great whale species are currently listed as endangered, some critically so, while a number of localized populations are gone or just about gone. The right whale of the North Atlantic, once common, is now down to a population of around three hundred and still declining. These giant animals were given their name in the old days because, as slow-moving and naturally-buoyant-after-death creatures, they were the "right" ones to find and kill. Now they are right for extinction, with their continuing decline today largely the consequence of accidental collisions with ships. The magnificent blue whales of the Antarctic, abundant until whalers discovered them, have been reduced to around 1 percent of their original numbers. At more than a hundred feet long and 150 tons heavy, incidentally, these animals are the largest creatures ever to have lived on this planet, land or sea, but they, too, are teetering on the edge of nonexistence. Also endangered or threatened are the gray whales of the northwestern Pacific, the fin whales, the sei, the beluga, and the sperm whales. "Trusting creatures whose size probably precluded a knowledge of fear," writes author and marine wildlife expert Richard Ellis, in *The Empty Ocean* (2003), "the whales were chased until they were exhausted and then stabbed and blown up; their babies were slaughtered; their numbers were halved and halved again." The industrialized killing of whales during the twentieth century, Ellis concludes, was "perhaps the most callous demonstration history offers of humankind's self-appointed dominion over animals. One searches almost in vain for an expression of sympathy, compassion, understanding, or rationality. In their place were only insensitivity and avarice."[6]

The International Whaling Commission (IWC) was originally organized in 1946 to support the industry by promoting the supposedly "sustainable" harvesting of whales. But commercial whaling had, by the second half of the century, reduced the numbers of most species so decisively that in July of 1982 the IWC declared a moratorium on all whaling.

That important and positive event has been challenged continuously by the Starbucks of this world. The Soviet whaling industry simply continued harvesting whales of all species, all ages and sizes, while falsifying their reports.[7] The Japanese officially adhered to the terms of the moratorium by identifying their whaling as a "scientific" rather than a commerical enterprise.[8] Iceland ignored the moratorium, allowing its ships to kill one hundred minke whales and as many as one hundred and fifty endangered fin whales during the 2008 and 2009 season. The Norwegians have never stopped whaling either and continue to slaughter hundreds of minke whales yearly, insisting that whale killing is a glorious part of their cultural heritage and, furthermore, that these giant mammals eat too many fish.[9] During the 2009 meeting of the IWC, meanwhile, Greenland, backed by the Danish government, applied for permission to harvest as many as fifty endangered humpback whales over the next five years for the purposes of "aboriginal subsistence," even though Greenland already has a surplus of whale meat, which is sold in supermakets.[10]

Whales are magnificent creatures who inspire us with their mass and power, whose continued existence lifts our spirits, and who remind us, in ways that few other living animals can, of our own fragile impermanence. What a depressingly impoverished world we and our children and their children stand to inherit, as we continue to wipe out some of the grandest animals ever to have lived, all done in the name of commerce and pragmatism, all carried out in the Starbuckian pursuit of a dollars-and-cents profit.

The vision of Starbuck as an average or normal man leads us to wonder whether there is any hope at all. If the norm of our species is essentially Starbuckian, can this war against nature and animals ever be ended or even remotely resolved? Are we bound by our very nature, our human nature, to destroy the biosphere, the living world, piece by piece, species by species, and ultimately, then, to destroy ourselves? Or can we hope for something someday that resembles a truce with nature and animals, an actual time of peace? Is there any hope?

The historical record so far is not encouraging. All the great whales were decimated in the instant, almost, that people had the technological capacity to do it; and that grim pattern seems to have been repeated throughout history and back into prehistory. Some paleohistorians credit the arrival of nomadic bands of human hunters carrying an improved stone tool technology for the so-called Pleistocene Die-off: the extermination of much of the great variety of larger and very large terrestrial species alive during the recent Ice Age. Almost three quarters of all North American mammals more than forty kilograms in weight—around forty species altogether—were extinguished quite suddenly (within a few thousand years) during the late Pleistocene. Three quarters of all similarly large animals living in South America also went extinct in the same period, while in Australia, nine tenths of all the large animal species disappeared.[11]

Three remaining land giants from the Pleistocene, the three species of modern elephants living in Africa and southern Asia, found their refuge in hot, harsh environments with relatively low human numbers. But then, suddenly, humans appeared with new projectile technologies: simple rifles, at first, followed by military-style submachine guns. The result? The world elephant herd has been cut in half during the last thirty years.

Most animals living in Africa's Congo Basin forests were until recently protected by the sheer size and complexity of this rain-forest system, the world's second largest. But during the last twenty years, people riding large bulldozers and wielding powerful chain saws have broken into the protective vault of these forests. International logging companies serving European and Asian consumers have arrived, built roads, and begun cutting down trees, some of which were contemporaries of Michelangelo. The loggers, meanwhile, continue to promote their activity as "sustainable utilization" of forests for the sacred purpose of human "development." The result? Virtually every animal species in the Congo Basin is now under assault from an army of commercial meat hunters, who move along the logging roads and currently remove somewhere from one to five million metric tons of animal flesh each year, much of that going to urban markets to be sold as a luxury fare, at prices higher than those for comparable cuts of domestic animal meat. Among those many forest species are all three African great apes: chimpanzees, bonobos, and gorillas. All three are going down.

That's what I mean when I speak of the war against animals. It's the story I've been talking about, the same old tale of mad Ahab and sane Starbuck rampaging through the natural world, and it has been repeated again and again and again. Different details, same pattern. As soon as humans have the technology to destroy nature—a species, a group of species, an ecosystem, a living world—they have done so. It is easy enough to imagine that the end of this story will occur only after the destructive processes we have set in motion turn back on us.

Early in this book, I introduced the concept of Darwinian narcissism, which is the evolved inclination of all organisms to orient themselves to their own kind. Darwinian narcissism among humans, so I suggested then, can be described on a number of levels—emotional, psychological, perceptual, intellectual, moral, and so on. Emotionally you and I are most fully attached to other humans. Psychologically, we recognize ourselves as being human, and we continually seek the company of other humans to interact with, to love, hate, emulate, scorn, or studiously be indifferent to. Morally, we see ourselves as dealing almost wholly with each other, as we negotiate, through our moral systems, the inevitable conflict between self and others.

Some people like to speak of the problem of *speciesism*—a recently coined word that intentionally evokes an analogy with racism and sexism.[12] If we can just overcome this latest ism, so the thinking goes, we will have solved the problem and ended the war against animals. Unfortunately, however, that analogy is misleading. Since both racism and sexism have to do with how we treat members of our own species, they can directly be overcome by appeals to fairness and empathy and other elements of our common moral understanding. We humans already possess all the psychological equipment needed to respond positively to those two negative isms. But the problems associated with the third ism, this speciesism, belong in a completely different category of moral debate. No mother will ever trade the life of her child for the life of a laboratory rat, no matter how many earnest slogans are shouted, no many how many exhortative placards are waved in her face. She won't make the trade because it would fundamentally violate her deepest sense of things, her own human nature. The heart, the human heart—by which I mean the inherited emotional systems that define human nature—tells us that our main orientation is always to other members of our own kind. By our very nature we are powerfully inclined to see and seek the world in the face that is our own.

Human nature is what we are given and who we are. But human nature is a lot more malleable than we may at first imagine. We cannot change it fundamentally, but we can modify it tremendously—and the best evidence for that assertion could be your case or mine.

Start with mine. I am just an ordinary person, one who lives in a house with other people, takes vacations with other people, falls in love with other people, has sex and produces children, human children: an ordinary person fully caught up in his own drama as shaped by the text of Darwinian narcissism. If I go to the park with my two dogs and let them run around with other dogs, I prefer talking to the dog owners. I won't get down on all fours and commune woofily and sniffily with the dogs, who have their own world to be narcissistically engaged in. At the same time, though, I'm fascinated by animals, and I care deeply about animals of all kinds. I even suspect I have moral feelings about them. How did that happen? Why should that be? Why should I care?

During the last several years, I've spent some time investigating elephant meat and ivory smuggling in Central Africa, talking with smugglers and transporters and marketers. I've also had some interesting conversations with gorilla hunters working the forests of Central Africa, as well as with others involved in the illegal trade in ape meat as it moves into the big cities. In northeastern Burma, I've met people involved in the illegal trade in endangered-animal parts that are abundantly passed across the border into China, serving the whims of an increasingly affluent Chinese middle class. I've had many informative chats with people in many places who sell or use animals, live and dead animals of any kind, for any purpose you might imagine.

When I ask this one about the endangered bears he keeps in cages and feeds a bucket of slop to each day, animals he is planning soon to chop up and pass across the border into China, he explains his plans with eagerness. It's how he hopes to make money, big money. When, standing in the middle of an African forest, I ask that one about the endangered gorillas he has shot, ruthlessly and methodically, as a profit-making business activity, he describes his killing methods as if they were among the most mundane activities in the world. "God gave us gorillas to eat," he explains simply. When I ask another person about the dried and fly-infested elephant meat she is trying to sell in a market in Bangui, capital city of the Central African Republic, she displays more of the meat and extolls its many great virtues. These, in my experience, are all perfectly decent, interesting, and ordinary people. They are merely trying to make a living, doing what they have learned to do. They hope to survive and to seek all the good things offered by life. In that way, they are very much like you and me, yet they do not really seem to care about animals at all—at least not in the way I do.

Why? Or rather, what's the difference between them and me? I claim to be ordinary as well. I like these people and have felt genuine pleasure in meeting them. So how have we come to feel so differently on this issue? Why do we hold such powerfully contrasting viewpoints on the morality of our treatment of animals—or rather, why should I be the one who thinks our treatment of animals might sometimes be described in moral terms? In this book, I've been insisting that morality is largely limited by the constraints of Darwinian narcissism, and I could present these people as prime examples. But how is it possible that I also seem to see morality, human morality, as flexible enough, sometimes, to reach beyond those constraints? What's the difference between these people and me?

One difference, you might think, is economic circumstance. None of them is starving, but the people I've been speaking of—including the bear farmer in northeastern Burma, the gorilla hunter and elephant-meat marketer in Central Africa—have at best only a few choices about how they make a living. The difference between their economic condition and mine is obviously important, perhaps overwhelmingly so. I don't want to minimize it. However, it doesn't really explain much. Plenty of rich people have absolutely no feelings about animals at all. I've met Americans, people living in one of the world's richest economies, well educated and quite well-off, who also clearly and routinely abuse animals; and I've met other rich or well-off people who are probably involved in the illegal smuggling of endangered species. Come to think of it, some entirely wealthy men spend large sums of money for the excitement of shooting holes into elephants in Africa or polar bears in the Arctic. So different economic circumstances don't explain much about different attitudes about animals.

A second difference, some cynics like to say, is intelligence or sophistication. Those who care about animals, these cynics will tell you, are just naive or simpleminded, or simply untraveled or inexperienced. They're weak in the head, the sort who see every deer as Bambi. But level of intelligence or sophistication doesn't seem to be much of a reason for the difference in attitudes either, since I know some patently smart, accomplished, and widely traveled people who also are vegetarians for, they say, ethical reasons. They care about animals rather in the way I do.

No, I think the best evidence will tell us that caring about animals is not a direct consequence of wealth or of weak-mindedness, but is, rather, a simple learned condition, one acquired, as most learned conditions are, through experience, education, and cultural tradition. If Darwinian narcissism is a matter of human nature, then caring about animals is a case of human nurture interacting with that nature.

Cultural traditions offer the clearest evidence of how cleanly, even precisely, our feelings about animals can be acquired, because cultural traditions produce thousands of oddly variant stances that sometimes look completely arbitrary. It's a cultural tradition in my part of the world to tolerate and often to love dogs. I love dogs. Half the people I know do, including my friend the hunter, who seriously grieved at the death of his own dog but who also, because he's grown up in a hunting tradition, likes to trap coyotes. A dog is genetically a wolf made paedomorphic through domestication. The difference between a paedomorphic wolf and a coyote is not much, in my estimation. So this killing of coyotes is where he and I differ, but I predict we would be equally distressed were we jointly to visit the Live Animal Restaurants Street of Mong La, in Burma's Special Region 4, where cute little white puppies are on the menu and displayed live, alongside just about any other kind of animal you could imagine. The chef kills to order.

Yet if experience, education, and cultural tradition answer the question of why some people care about animals and others don't, even down to the level of particular species, we are still left with the following additional question: About what? Experience about what? Education about what? Cultural tradition about what?

The answer, I believe, can be summarized as experience, education, or cultural tradition about the psychological presence of animals. *Psycho-*

logical presence is another way of saying *mind* or of speaking about the capacity for *subjective mental experience*. That's the phrase I underhandedly borrowed from philosopher Thomas Nagel and used in chapter 1, when I was talking about the Third Way of Thinking about animals in light of Nagel's essay "What Is It Like to Be a Bat?"

Simply put, the Third Way of Thinking goes like this: This world is filled with mindful beings, most of them of the nonhuman variety. Many of the world's nonhuman minds could be altogether a lot simpler than mine, and a few could be, altogether, a good deal more complex and clever than I can ever really imagine or appreciate. But in any case they are, inevitably, alien minds, ones that will always be imperfectly comprehensible to me. This Third Way of Thinking about animals is something I've arrived at through experience, education, and cultural tradition, and the experience part is perhaps simplest to describe.

Yesterday, I visited my nephew Andrew at his workshop, which is also an airplane hangar. He had recently acquired a couple of barn cats, still kittens, to help with the mice problem, and so when I walked into the hangar and workshop, I was greeted by a couple of tiny creatures who instantly, upon seeing me, sidled over right in front of my feet, one kitten for each foot, looked up, way, way up, into my face and eyes, met my gaze with theirs, and mewed. These were kittens, let me repeat, which means they were both very small, and I would be, from their perspective, mountainous. Their body weight combined is roughly 1 or 2 percent of mine, and that makes them about as small, compared to me, as I would be next to a medium-size whale. That's a major size difference, yet somehow they appeared to have made some quick and interesting conclusions about me as I walked into the hangar and met them for the first time.

First, they were instantly able to distinguish me, my moving presence, from the nonmoving background, which looks like an impressive feat of cognitive processing. Second, they appeared to conclude that I was alive, not, say, a moving tractor or some other machine, and would therefore respond with some singular particularity to them. Third, they seemed to understand that I was the sort of living creature who could safely be approached. Fourth, they seemed to recognize that the center of my attention, or the key to my potential responsiveness, was located not at my feet but at my face and eyes. Fifth, they appeared to understand that I could fullfil their needs for closeness and stroking rather in the way that a mother cat would. If they mewed enough, I might actually pick them up (one at a time), cradle them, and stroke them. I did. They seemed to like it.

Now, you could argue that these cats have not at all demonstrated mindfulness. They have merely demonstrated simple reflex, raw instinct, and a bit of stimulus-response association, and so they are merely cute little biological machines I mistook for animals with minds. But we all know, deeply if intuitively, the difference between an animal and a machine. A machine irritates you when it breaks down. An animal—at least an animal you've spent meaningful time with, a pet, a domestic dog or cat—can cause you grief when he or she dies. As long as we are able to distinguish one from the other, no machine will arouse in us the same emotional response that a real dog can. (This, by the way, isn't an argument for vitalism, the concept of a force distinctive from physical or chemical forces that can distinguish animal from machine. If you grant that a computer-based robot of enough complexity could develop a mind and emotions, then a robot that's a perfect or near perfect imitation of a dog, with a dog's mind and emotions, ought to be experienced as one.)

Scientists interested in the subject have, generally, stopped asking whether animals have minds and begun concentrating on the problem of what kinds of minds. This is the question of animal intelligence or cognition, and the answer, obviously, varies widely from one species to the next. Some animals are pretty simple; others are quite complex—unpredictably and even creatively so. Some have remarkable abilities in areas of perception that are so far removed from ours we have trouble even conceptualizing their experience. Other animals are capable of learning to communicate through a humanlike symbolic language, or at least a provocative simulation thereof. Still other animals are cognitively complex enough to recognize themselves as individuals. Should they come across a mirror or some other reflective device, they may stop and preen themselves or examine parts of their body they cannot otherwise examine, which suggests not merely awareness but individual self-awareness. Some animals—including elephants and whales—have brains that look to be approximately as physically complex as ours and yet are significantly larger. Others have brains that may be larger or smaller than ours, yet begin to approach the human ratio of brain to body size.

Cetaceans—whales, dolphins, and porpoises—are of particular interest here because they have enormous brains, and we are logically obliged to ask why such big brains evolved. A sperm whale's brain is two thirds again the size of an elephant's, while the brains of all toothed whales, including dolphins, are bigger than any nonhuman ape's and, in proportion to body size, come in second only to humans'.[13]

Cetaceans have been evolving separately from primates for more than ninety-five million years, so we might speculate that the large brains of cetaceans have little relationship to intelligence of the sort that humans and other primates show; but a number of experts think otherwise. In spite of such an extended period of separate evolution, these experts conclude, the expanded size of the "insular and cingulate cortices in cetaceans is consistent with high-level cognitive functions—such as attention, judgment, intuition, and social awareness—known to be associated with these regions in primates."[14] The largest brains among the cetaceans are, moreover, filled with a large number of specialized *spindle* neurons, associated with complex social cognition, which were until recently considered unique to the brains of humans and the great apes. Altogether, the brains of cetaceans show "a structural complexity that could support complex information processing, allowing for intelligent, rational behavior."[15]

That's the word from the neuroanatomists, and those scientists who have been studying cetacean behavior for the last several decades strongly support it, finding a group of animals who have excellent memories and high levels of social and self-awareness, who are excellent at mimicking the behavior of others and can respond to symbolic representations, who form complex and creatively adaptive social systems, who show a broad capacity for the cultural transmission of learned behaviors . . . and so on. These and other abilities vary widely from one species to the next, not surprisingly, but the species most fully studied—bottlenose dolphins, killer whales, humpback whales, sperm whales—are impressively sentient and creative. They're highly intelligent animals, and it's clear to me that Herman Melville suspected as much.

What makes Melville's novel *Moby-Dick* unique, in the library of English-language literary classics, is the presence of an intelligent animal protagonist: a mind in the waters. Perhaps the closest thing we have to this book would only begin to appear later on with the emergence of an increasingly fertile science fiction where humans experience an invigorating if often disastrous contact with extraterrestrial beings of some intelligence: minds from outer space.

At this point, such science fiction could be moving closer to science fact. It's reasonable to imagine that life may have evolved in other places in the universe. Astronomers have recently identified some 370 *exoplanets*, a fancy word that describes planets orbiting stars other than our own sun. That may seem a small number, and generally these known exoplanets are inhospitably hot or cold or otherwise unsuitable for the appearance of life. But since that sample of 370 exoplanets—found within four hundred light-years of our own solar system—represents the tinest fraction of possibilities in a universe with trillions of suns that might include planets, we should recognize it as a number with great promise. There could be many billions of yet undiscovered planets in the Milky Way galaxy alone, which contains two hundred billion stars, give or take a few.[16] Thus, it is not unreasonable to expect that somewhere out there—or in many places—life has already emerged as it did on our planet. Perhaps, somewhere out there, life has evolved in such a way and for a long enough time to produce intelligent beings, extraterrestrial creatures with minds.

Not long ago, I would have been taking a risk merely speaking about these ideas. The Italian philospher Giordano Bruno was burned alive in 1600 for declaring that intelligent life might exist in another world, and of course Galileo was tried for heresy in 1633 and forced to announce he had changed his mind about the Earth turning around the sun. These days, however, Galileo's erstwhile persecutors are actually quite open to the possibility of extraterrestrial intelligence. In the fall of 2009, the Vatican's astronomer, the Reverend José Gabriel Funes, hosted a five-day conference of astronomers, physicists, biologists, and religious ethicists to consider the matter of "whether sentient life-forms exist on other worlds."[17] That's an important shift in position for a major world religion, and I wish I had been invited to the conference, since I would have liked to ask the following question: If all of this science fantasy should turn into science fact, would we really welcome contact with mindful beings from outer space?[18]

If we reached them before they reached us, we would do so probably because we're technologically more advanced than they. If that's the case, what indicates that we'll treat them any differently from the way we've treated every other creature with sentience already here on this planet? We know elephants are intelligent and highly emotional. They have been shown to recognize themselves as individuals in a mirror, which means they are self-aware or self-conscious. They also understand one another as individuals. They have excellent memories and a strong sense of empathy. Yet we continue to harvest them, both legally and illegally, because some people find their front teeth aesthetically pleasing when carved.

And if the extraterrestrials should reach us first because they're technologically more advanced than we, what in our experience suggests that their high intelligence and great technology would be accompanied by great wisdom and kindness or would automatically promote high sympathy for another species? It seems likely that they will have emerged through the same basic evolutionary process that produced us, one involving speciation and, after the division into species, inevitably producing what I call Darwinian narcissism. We should really be worried that these intelligent and technologically advanced visitors from outer space will take one look at us and decide our teeth would make good trinkets, or that our brains, when processed, will provide just the right kind of lubricant for some arcane purpose only they understand.

Perhaps the best hope we have is to demonstrate with our own example that, given time, we can achieve a greater wisdom about ourselves and our relationship with the rest of the natural world down here, on this planet. To show that we can find common purpose, in our relationship with other biological beings, and thereby discover a more fulfilling sense of ourselves and others and, finally, a peace with animals and nature. . . . Ohhhh, you're absolutely right: Peace is a suspiciously remote and ethereal concept, one far too abstract to end a book with. But such a peace can, I think, come as the steady accumulation of particularized moments: the moment you looked and wondered but did not take, the moment you feared but did not destroy. Peace of the sort that comes from the wisdom of seeing something you do not fully understand, something that could be dangerous or maybe not, but that you decide in any case to leave alone.

Before she studied elephants, bioacoustician Katy Payne applied her scientific training and observation skills to whales, listening for fifteen years to the extended, complex, and perpetually changing songs of humpback whales.

One time, while scouting out locations on the northeastern coast of Argentina's Valdés Peninsula to begin new research on the behavior of southern right whales, Payne heard of a place where the ocean depth and currents conveniently drew whales near to shore. Indeed, as soon as she and a friend, Ollie Brazier, launched their small boat at that same spot, they watched a whale drift right past them. They maneuvered into a position where they, too, were drifting in the current, and then they cut the engine.

The giant creature ahead of them turned around and swam their way, disappearing beneath the surface, soon to emerge as a glistening wall rising up alongside the boat, drifting there for about two minutes. The wall was the underside of the chin of a vertical whale. A pair of eyes, located on either side of the chin, were just below the surface and apparently examining the boat.

The wall slipped back to the horizontal, and the whale once again turned downstream in the current, ahead of the boat, but now he began swimming backward, back in the direction of the boat. Payne and Brazier could recognize the creature's underwater presence by a ruffled flurry on the surface, and then they saw, right beneath the surface, a massive tail waving slowly back and forth. The tail-waving may have been a threat, and certainly that immense and gracefully flexible tail could have raised itself out of the water, covered the boat twice over, and crashed down on it and the people inside. Instead, however, the enormous appendage simply flattened out and reached, as if it were the open palm of a hand, right beneath the boat.

With his tail thus flattened, the whale lifted the boat entirely clear of the water and held it and the two people in it above the surface for a minute. "He held us steady for a full minute," Payne writes, "two people on a tray six inches above the water's surface."[19]

The whale then lowered them and their craft gently back down. Payne looked into the dark water and saw the giant mammal swim or drift downstream again, once more drawing ahead of the moving boat—and then she saw him once more swim in reverse. She saw the tail wave back and forth, beneath the surface, and then she saw it again reach out flatly, beneath their small boat, and again the vessel and its two astonished occupants were lifted entirely out of the water. After a time, the tail lowered them gently back onto the water. The whale swam or drifted forward in the current, then again swam in reverse. And a third time the vast creature made the same gesturing wave, the same deft reach, the same gentle lifting of vessel and occupants, the same gentle lowering.[20]

It's hard to know what the whale was thinking or experiencing that day as he surveyed two alien beings in an alien vessel, gently measured their heft and probed their significance before leaving them intact; but it's easy to believe that he had thoughts and a subjective mental experience. And it may be easy enough to conclude that he examined, considered, and, with some degree of deliberation, chose not to destroy what he did not entirely understand. That's what I mean by a particularized moment of peace.

Acknowledgments

This book began as an argument at a dinner party hosted by my friend Ajume Wingo, and I profoundly appreciate Ajume's strong and repeatedly expressed enthusiasm for the core idea.

I am also grateful to some other close friends, especially Jane Goodall and Sy Montgomery for their warm encouragement, and Marc Bekoff, Daniel Dennett, and Richard Wrangham for their intelligent and extensive criticisms after reading the typescript in various stages. I must express my intellectual indebtedness to two pioneering experts who have already written thoroughly and excellently on the subject of animal social behavior and the evolution of morality: Marc Bekoff and Frans de Waal. De Waal has traveled widely in these regions, and I have significantly relied on his work on empathy as well as some of his other distinctive ideas. Other experts who generously answered questions and pointed in useful directions include Richard Connor, Bella DePaulo, Takayoshi Kano, Mark Laidre, Katy Payne, Michael Poole, David Reier, and Laurie Santos. For guidance of a more pragmatic sort, I thank Karl Ammann, Kathy Ammann, Myo Minn Aung, Soe Min Aung, Iain Douglas-Hamilton, Josiane Dwili, Mbongo George, Darra Goldstein, Melissa Groo, Ko Gyi, San Lwin, Stephanie Magba, Joseph Melloh, Rita Mesquita, Marion Meyer, U Tun Nyan, John Oates, David Oren, Pedro Pimentel, Ron Pontier, Karlo Saddeau, Moses ole Sipanta, Andrea Turkalo, and Sai Win.

Peter Ginna, my editor at Bloomsbury, has most kindly welcomed me into the fold, as have Pete Beatty and Laura Phillips, also with Bloomsbury. I thank them. My two grown children, Britt and Bayne, have assisted as informed readers and critics, for which I am likewise deeply thankful. And my wife, Wyn Kelley, who has her own career teaching literature at the Massachusetts Institute of Technology, has beautifully served as both my first editor and my favorite expert on Herman Melville's *Moby-Dick*. Finally, I should acknowledge the long-term contributions of three men who have in various ways, and at various critical times in my life, served as teachers, guides, and mentors: Jarold Ramsey, Thomas Moser, and Peter Matson. To those three this book is dedicated.

Notes

1. Words

1 Melville, [1851] 2002, p. 139.

2 Douglas-Hamilton and Douglas-Hamilton, 1975, pp. 102–3.

3 Chadwick, 1992, pp. 254–57.

4 Ibid., p. 256.

5 Ibid., p. 257.

6 Sikes, 1971; Shoshani and Eisenberg, 1982.

7 Sukumar, 2003, p. 149.

8 Payne, 1998; Payne, Langbauer, and Thomas, 1986.

9 Weissengrubber et al., 2006; O'Connell, 2007.

10 Chadwick, 1992, p. 257.

11 This story is mainly based on Scigliano, 2002, pp. 201–3. See also Alexander, 2000.

12 Price, 1992.

13 Scigliano, 2002, p. 204.

14 Alexander, 2000, p. 143.

15 Lewis, 1964, p. 102.

16 Ibid., p. 115.

17 Evans, n.d., pp. 144–45.

18 Evans covers several cases. Ibid., pp. 146–53.

19 Ibid., pp. 114–15.

20 Descartes, [1637] 1993.

21 Starbuck, interestingly enough, shares the inclination of all the other men

on this whaling ship to speak of whales with the masculine personal pronoun.

22 Darwin, [1859] 1996.

23 Darwin, [1872] 1998.

24 Darwin, 1871, p. 127.

25 Stanford, 2001.

26 Cheney and Seyfarth, 1990, provide a good review of these issues, pp. 1–13.

27 Griffin, [1976] 1981, p. 3.

28 Nagel, 1974.

29 Horowitz, 2009, p. 48.

30 Ibid., pp. 71–72.

2. Orientations

1 Melville, [1851] 2002, p. 20.

2 Lovejoy, 1980.

3 Peterson, 1989, pp. 26–29.

4 Myers, 1984, pp. 50–64; Caufield, 1985, pp. 59–61; Peterson, 1989, p. 34.

5 Rylands and Mittermeier, 1982; Thornback and Jenkins, 1982, pp. 161–62; Peterson, 1989, p. 35.

6 Peterson, 1989, pp. 24–26.

7 Wolfheim, 1983.

8 Peterson, 1989, pp. 49–73.

9 Perry and Manson, 2008, p. 1.

10 Ibid., p. 2.

11 Da Fonseca, 1985; De Assumpcao, 1983; Mittermeier, 1982; Mittermeier, 1987; Mittermeier et al., 1982; Thornback and Jenkins, 1982, pp. 181–82.

12 Bourne, 1974, pp. 71–75; Coimbra-Filho and Mittermeier, 1977; Dietz, 1984; Kavanagh, 1983, pp. 81–82; Mallinson, 1984; Thornback and Jenkins, 1982, pp. 137–47.

13 Johns, 1985; Johns and Ayres, 1987.

14 Kavanagh, 1972; Lippold, 1977; Van Peenen, Light, and Duncan, 1971.

15 Green and Minkowski, 1977; Karanth, 1985; Karr, 1973; Sugiyama, 1968.

16 Miller, 1903; Tenaza, 1975; Tenaza and Hamilton, 1971; Tenaza and Mitchell, 1985.

17 Smith, 2008; Bekoff, 2010, p. 13.

18 Kavanagh, 1983, pp. 26–27; Sussman, Richard, and Ravelojaona, 1985; Tattersall, 1982, pp. 1–35.

19 Jolly, 1987; Richard, 1982; Richard and Sussman, 1974; Tattersall, 1982, pp. 337–40; Simon and Geroudet, 1970, p. 193.

20 See Judson, 2008; also Epley and Whitchurch, 2008; Pronin, 2008; Sundström, 2008.

3. Definitions

1 Melville, [1851] 2002, p. 117.

2 Najafi, 2007, pp. 87–88.

3 Moore, 1903.

4 Curry, 2006; Walter, 2006.

5 Thierry, 2000, p. 61.

6 Boehm, 2008.

7 Pinker (Steven), 2008, p. 429.

8 Ibid., pp. 429–30; Rozin, 1996; Rozin, 1997.

9 Parker, 2003.

10 MacLean, 1990.

11 Douglas-Hamilton and Douglas-Hamilton, 1975, p. 234; Moss, 1988, p. 271; Poole, 1996, p. 95.

12 Grandin and Johnson, 2009, p. 5.

13 Panksepp and Burgdorf, 2003, p. 535.

14 Ibid.

15 Ibid., pp. 535–36.

16 Ibid., pp. 536–43; Panksepp, 2005.

17 Moffett, 2010, p. 9.

18 Hölldobler and Wilson, 2009, p. xx.

19 Moffett, 2010, p. 144.

20 Indeed, play may be a characteristic activity of most mammals, but play behavior is also found among marsupials, birds, turtles, lizards, fish, and invertebrates, according to Burghardt, 2005, p. 382.

21 Van Hooff and Preuschoft, 2003, p. 266.

22 Bekoff and Pierce, 2009, p. 116.

23 Panksepp, 2005.

24 Van Hooff and Preuschoft, 2003, p. 267.

25 Burghardt, 2005, p. 382, suggests five specific criteria.

26 Bekoff and Pierce, 2009, p. 119.

27 Ibid., p. 117; also Biben, 1998; Heinrich and Smokler, 1998; Miller and Byers, 1998; and Watson, 1998.

28 Fagen, 1981; Burghardt, 1998.

29 Bekoff and Pierce, 2009, pp. 123–24.

30 Watson, 1998, p. 67.

31 Van Hooff and Preuschoft, 2003, p. 268 (figure 10.4).

32 Bekoff and Pierce, 2009, p. 124.

33 Ibid., p. 123.

4. Structures

1 Melville, [1851] 2002, p. 140.

2 Weeks and Karas, 2006, p. 138.

3 Cudd, 2007; Rousseau, [1762] 1968.

4 Cudd, 2007; Boucher and Kelley, 1994; Gauthier, 1986; Narveson, 1988; Vallentyne, 1991.

5 Wiedenmayer, 1997.

6 Mineka and Cook, 1986; Cook and Mineka, 1989.

7 Wright, 1994.

8 Wrangham, 2009.

9 Ibid., pp. 1–6; also Klein, 1999; Lewin and Foley, 2004; Wolpoff, 1999.

10 Fouts and Mills, 1997; Gardner and Gardner, 1989; Miles, 1993; Patterson and Linden, 1981; Savage-Rumbaugh and Lewin, 1994.

11 Cheney and Seyfarth, 2007, p. 258.

12 Ibid.; also Kaminski, Call, and Fischer, 2004.

13 Pepperberg, 1999.

14 Cheney and Seyfarth, 2007, pp. 217–72.

15 Lieberman, 2007, p. 47.

16 MacAndrew, 2009, p. 1.

17 Shu et al., 2005, p. 9643.

18 MacAndrew, 2009; Enard et al., 2002.

19 Hauser, 2006, xvii.

20 In reference to Dugatkin, 2006.

21 Boehm, 2008.

22 Frank, 2005; Hodgson and Knudsen, 2006; Simon, 1990; Stark, 1961; Tenaka, 1996.

23 Hare et al., 2005.

24 Boehm, 2008, p. 332.

25 Ibid., p. 334.

26 Brosnam and de Waal, 2003.

27 Boehm, 2008, p. 335. See also Boehm, 1999.

28 Grandin and Johnson, 2005, p. 155.

29 McGlone, 2002.

30 "Ten Commandments," 2008.

5. Authority

1 Melville, [1851] 2002, p. 11.

2 Millan and Peltier, 2006, p. 3.

3 Grandin and Johnson, 2009, pp. 26–28; Mech, 1995; Mech, 2000.

4 Grandin and Johnson, 2009.

5 Millan and Peltier, 2006, pp. 68–74.

6 Whenever two or more individuals gather "in a group or relationship that involves, in whatever degree of informality or formality, the distribution of responsibilities, duties, needs, expectations, privileges, and rewards, a pattern of authority is present," according to Nisbet, Page, and Perrin, 1977, p. 106.

7 Grandin and Johnson, 2009.

8 Coppinger and Coppinger, 2004, pp. 63–67; Horowitz, 2009, pp. 35–37; also Belyaev, 1979; Hare et al., 2005; Trut, 1999.

9 Horowitz, 2009, p. 36.

10 Kistler, 2006, p. 127.

11 Gale, 1974, p. 109.

6. Violence

1 Melville, [1851] 2002, p. 242.

2 Goodall, 2001, p. 207.

3 Ibid., p. 216.

4 Ibid., p. 220.

5 Hrdy, 2009, pp. 3–4.

6 Wrangham and Riss, 1990.

7 Hiraiwa-Hasegawa, 1999.

8 Arcadi and Wrangham, 1999; Murray, Wroblewski, and Pusey, 2007; Pusey et al., 2008; Sherrow and Amsler, 2007.

9 Lorenz, [1963] 1966, p. 129.

10 Rudolf and Antonovics, 2007.

11 Hiraiwa-Hasegawa, 1999, p. 326.

12 Harris, 1989, p. 428.

13 Ibid., pp. 432–33.

14 Elgar, 1992, p. 141.

15 Ibid., p. 140.

16 Baur, 1992; Mock, 1992.

17 Baur, 1992.

18 Peterson, 2003, pp. 80–103. See also, for example, Breman et al., 1999; Butinsky, 2000; Gao et al., 1999; Georges, 1999; and Hahn et al., 2000.

19 See Rudolf and Antonovics, 2007. Their conclusion of a likely "infrequent" disease transmission is based on the faulty assumption that the only transmission route would be from direct contact between eaten and eater during consumption. The assumption fails to consider examples of widespread infection through blood contact that can occur through means other than body to mouth. Ebola, for example, will spread rapidly and broadly during a case of predation (or cannibalism) even when only one individual actually consumes the meat.

20 Sussman, Garber, and Cheverud, 2005, p. 84.

21 Ibid., p. 90.

22 Silverberg and Gray, 1992; de Waal, 1992; Strayer, 1992.

23 Tucker, 2010; Packer, Scheel, and Pusey, 1990.

24 Packer et al., 1988.

25 Beatram, 1975; Packer and Pusey, 1983; Packer et al., 1988; Packer, Scheel, and Pusey, 1990.

26 Tucker, 2010, p. 30.

27 Kruuk, as quoted in Wrangham and Peterson, 1996, p. 287.

28 Frank, 1986; Kruuk, 1972; Wrangham and Peterson, 1996, pp. 153–55.

29 Wilson, Britton, and Franks, 2002; Wilson, Hauser, and Wrangham, 2001.

30 Wrangham, Wilson, and Muller, 2006; Wrangham and Wilson, 2004. Hunters have an annual risk of death from war of 0.164 percent, while farmers have an annual risk of 0.595 percent. The figures of human pre-state deaths come from a sample of 164 societies where the data were most readily available; the figures for chimpanzee deaths are taken from the five longest-studied populations with more than one community under observation.

31 Goodall, 1986, pp. 506–7.

32 Wilson and Wrangham, 2003.

33 See, for example, Wilson, Wallauer, and Pusey, 2004.

34 Power, 1991, p. 241.

35 Sussman and Marshack, 2010.

36 Ibid.

37 Ibid.

38 Smith, 2007, p. 18.

39 Ibid.

40 Bronner, 2009.

7. Sex

1 Melville, [1851] 2002, p. 306.

2 Kano, 1992, pp. 26–34; de Waal, 1997, pp. 23–34; Wrangham and Peterson, 1996, pp. 200–204.

3 De Waal, 1988.

4 Kano, 1979, p. 130.

5 Wrangham, McGrew, and DeWaal, 1994.

6 Idani, 1991a; Wrangham and Peterson, 1996, p. 214.

7 De Waal, 1997, p. 29, including quote.

8 Kano, 1992, p. 140.

9 Ibid., p. 141; Savage-Rumbaugh and Wilkerson, 1978; Patterson, 1979.

10 Savage-Rumbaugh and Wilkerson, 1978, p. 337.

11 Kano, 1992, p. 178.

12 Roughgarden, 2004, p. 137; Bagemihl, 1999, pp. 269–476.

13 Roughgarden, 2004, p. 136; Bagemihl, 1999, pp. 479–655.

14 MacFarquhar, 2008.

15 Human sexual diversity: one of the critical arguments in Roughgarden, 2004.

16 See Leiber, 2006.

17 Sharp, 2001; Abrams, 2006; "Polygamy," 2006.

18 Whyte, 1978, p. 222.

19 Goodall, 1986, pp. 443–87.

20 Ibid., pp. 453–65.

21 Ibid., pp. 466–67.

22 Ibid., p. 466.

23 Ibid., p. 467.

24 Furuichi, 1989; Idani, 1991b; Wrangham and Peterson, 1996, p. 209.

25 Fossey, 1983.

26 Fisher, 2004, p. xii.

27 Tucker, 2010, p. 33.

28 Fisher, 2004, p. 49.

29 Ibid., p. 50.

30 Roughgarden, 2004, p. 54.

31 Otter and Ratcliffe, 1996; Roughgarden, 2004, p. 55; Smith, 1991.

32 Roughgarden, 2004, p. 57.

33 Peterson, 1989, pp. 204–6; Tenaza, 1975; Tenaza and Hamilton, 1971.

8. Possession

1 Melville, [1851] 2002, p. 308.

2 Wrangham, 2000, p. 34.

3 Peterson and Goodall, 1993, pp. 41–48; also Boesch and Boesch-Achermann, 1981; Boesch and Boesch-Achermann, 2000.

4 Goodall, 1986, p. 374; for more general issues of food stealing, begging, and sharing, see pp. 272–96, 372–76. See also Goodall, 1968; Teleki, 1973.

5 Goodall, 1986, p. 299; Wrangham, 1975.

6 Raby et al., 2007.

7 Ibid.

8 Dally, 2007; also Emery and Clayton, 2001.

9 Dally, 2007.

10 Ibid.

11 Ibid.; Dally, Emery, and Clayton, 2006.

12 Dally, Emery, and Clayton, 2006.

13 Dally, 2007; Emery and Clayton, 2001.

14 Begging of Humphrey, Jomeo, and Figan: Goodall, 1986, p. 304; stealing by Satan: Goodall, 1986, p. 303.

15 Wilson, [1975] 2000, p. 332; Wiley, 1973; Wiley, 1974.

16 Wilson, [1975] 2000, pp. 368–71.

17 Kummer, 1968; Kummer, 1995.

9. Communication

1 Melville, [1851] 2002, p. 53.

2 Lloyd, 1965.

3 Santos, Nissen, and Ferrugia, 2006; Gyger and Marler, 1988.

4 Bond and Robinson, 2005; also Breed, 2001.

5 DePaulo, 2004; Angier, 2008.

6 Bok, 1978, p. 19.

7 Whiting, Webb, and Keogh, 2009; also Skilton, 2009; Whiting, 2000; and Zuckerman, 2009.

8 Kluger and Masters, 2006.

9 Ekman, [1985] 2009.

10 Dennett, 1983; Bryne and Whiten, 1987; Lewin, 1987. See also Savage-Rumbaugh and Lewin, 1994, p. 272; Flombaum and Santos, 2005; Santos, Nissen, and Ferrugia, 2006.

11 Cheney and Seyfarth, 2007, p. 167; also Bräuer et al., 2006.

12 Cheney and Seyfarth, 2007, pp. 165–66.

13 De Waal, 1982, p. 49.

14 Ibid.

15 Ibid., p. 133.

16 Ibid., p. 74.

17 Goodall, 1986, p. 579.

18 Ibid., p. 577

19 Ibid., p. 581.

20 Ibid., p. 582.

21 Woodruff and Premack, 1979.

22 Fouts and Mills, 1997, p. 46.

23 Ibid., p. 156.

24 Plooij, 2000, p. 88.

25 Ibid.

10. Cooperation

1 Melville, [1851] 2002, pp. 75–76.

2 Drea and Frank, 2003.

3 Dugatin, 1999; Slater, Schaffner, and Aureli, 2007.

4 Goodall, 1986, p. 357.

5 Bekoff and Pierce, 2009, p. 3.

6 For example, Pfaff, 2007.

7 Darwin, 1871, pp. 71–72.

8 Huxley, [1894] 2009.

9 Nowak et al., 2006.

10 Porat and Chadwick-Furman, 2004.

11 Connolly and Martlew, 1999, p. 10.

12 Sherman, 1977.

13 Connolly and Martlew, 1999, p. 10.

14 Fisher, 1930; Haldane, 1955; Hamilton, 1963; Hamilton, 1964; Smith, 1964.

15 Boesch, 2003; Boesch, 1994.

16 Wilkinson, 1988; Wilkinson, 1984.

17 Axelrod and Hamilton, 1981.

18 Johnson, Stopka, and Macdonald, 2003; Barrett et al., 1999.

19 Hart and Hart, 1992.

20 Trivers, 1971; also Taylor and McGuire, 1988.

21 Heinsohn and Packer, 1995; Packer, 1988.

22 Rilling et al., 2002; Watson and Platt, 2006; Decety et al., 2004.

23 Stevens and Hauser, 2004.

24 Brosnan and de Waal, 2003.

25 Rutte and Taborsky, 2009; Brosnan, Schiff, and de Waal, 2005.

26 Silk, Seyfarth, and Cheney, 1999, p. 689.

27 Kosfeld et al., 2005.

11. Kindness

1 Melville, [1851] 2002, p. 255.

2 Payne, 2003, p. 82.

3 See, for example, Hoffman, 1981.

4 Poole, 1996, 162–63.

5 de Waal, 2005, p. 2.

6 Bekoff and Pierce, 2009, p. 1; also de Waal, 2005, p. 2.

7 O'Connell, 1995, p. 403.

8 Goodall, 1986, p. 378.

9 O'Connell, 1995.

10 Flombaum and Santos, 2005; Santos, Nissen, and Ferrugia, 2006.

11 O'Connell, 1995, p. 401; Savage-Rumbauch and McDonald, 1988, p. 228.

12 Dennett, 1988.

13 O'Connell, 1995.

14 Preston and de Waal, 2002.

15 Sagi and Hoffman, 1976; Ungerer et al., 1990; Zahn-Waxler et al., 1992; Zahn-Waxler, Friedman, and Cummings, 1983; Zahn-Waxler, Robinson, and Emde, 1992.

16 Bekoff and Pierce, 2009, p. 96; Balcombe, Barnard, and Sandusky, 2004.

17 Langford et al., 2006, p. 1967.

18 Ibid.; see Church, 1959; also Carey, 2006; Ganguli, 2009.

19 Based on Preston and de Waal, 2002.

20 Based on Blakslee, 2006.

21 Rizzolatti and Craighero, 2004.

22 Blakeslee, 2006.

23 Lyons, Santos, and Keil, 2006; Vedantam, 2006; Moll, Oliveira-Souza, and Zahn, 2008; Gallese, Ferrari, and Umilta (commentary in Preston and de Waal, 2002).

24 Ungerer et al., 1990, p. 94.

25 Quote from Rice and Gainer, 1962, p. 123.

26 Ibid.

27 Rice, 1964, p. 167.

28 Masserman, Wechkin, and Terris, 1964, p. 584.

29 Bates et al., 2008.

30 Kunz et al., 1994.

31 Dawkins, [1976] 2006, p. 4.

32 Preston and de Waal, 2002, p. 4; mine is a simplified version of their description of a "Perception-Action Model," where the "activation" of an "attended perception of the object's state, situation, and object . . . automatically primes or generates the associated autonomic and somatic responses, unless inhibited."

33 Anderson and Keitner (commentary in Preston and de Waal, 2002).

12. Duality

1 Melville, [1851] 2002, p. 307.

2 "Trouser-Wearing Woman," 2009.

3 Fisher, 1992, p. 281.

4 Lerner, 1986, p. 22.

5 Wrangham and Peterson, 1996, pp. 119–20. For example, Mead, 1949, p. x; Fisher, 1992, p. 283; Lerner, 1986, p. 31.

6 Thomas, [1959] 1968.

7 Thomas, personal communication.

8 Wrangham and Peterson, 1996, pp. 76–77; Daly and Wilson, 1988.

9 Lee, 1982, p. 45.

10 Lee, 1979.

11 Shostak, 1981, p. 246.

12 Ibid., p. 313.

13 Wrangham and Peterson, 1996, p. 121.

14 Freud, [1925] 1953, pp. 257–58.

15 Piaget et al., 1932, p. 77.

16 Haan, 1975; Holstein, 1976.

17 Gilligan, [1982] 1993, p. 69.

18 Ibid., p. 72.

19 Ibid., p. 104.

20 Ibid., p. xix.

21 See Pinker (Susan), 2008, for an informed response to this common if misguided notion, especially pp. 92–125.

22 Poole, 1996, p. 54.

23 According to Payne, 1998, p. 75.

24 Ibid.

25 Ibid., p. 96.

26 Ibid., p. 100.

27 Whitehead, 2003, p. 448.

28 Mesnick et al., 2003.

29 Whitehead, 2003, p. 449.

30 Harris, 1989, p. 278; Wrangham and Peterson, 1996, p. 111.

31 Strier, 1992a; Strier, 1992b; Wrangham and Peterson, 1996, pp. 174–76.

32 Wrangham and Peterson, 1996, p. 126n.

33 Quoted in Sacks, 2009.

34 Douglas-Hamilton and Douglas-Hamilton, 1975, pp. 53–54.

35 Chadwick, 1992, pp. 430–36.

36 Slowtow, Balfour, and Howison, 2001; Slowtow and van Dyk, 2001.

37 Siebert, 2006.

38 Bradshaw et al., 2005.

39 Seibert, 2006, p. 44.

40 Ibid., p. 47.

41 Bradshaw et al., 2005; also Owens and Owens, 2005; Slotow et al., 2000.

13. Flexibility

1 Melville, [1851] 2002, p. 406.

2 Miller, 1977; Finkelman, 2001; Helo and Onuf, 2003.

3 Ross, [1857] 2009, p. 27.

4 Perry and Manson, 2008, p. 137.

5 Ibid., pp. 39–40.

6 Ibid., p. 137.

7 Ibid.

8 Ibid., p. 138.

9 Ibid., pp. 251–59.

10 Smuts and Watanabe, 1990.

11 Perry and Manson, 2008, p. 140.

12 Kano, 1992, p. 185.

13 Hohmann and Fruth, 1993.

14 Kano, 1992, p. 176, table.

15 Ibid., p. 190.

16 Drea and Frank, 2003, p. 124.

17 Wrangham and Peterson, 1996, pp. 184–86.

18 Kano, 1992, pp. 185, 189.

19 Kaplan, 2006, p. 42.

20 Wrangham and Peterson, 1996, pp. 220–30.

21 "Women's Suffrage," 2009.

14. Peace

1 Melville, [1851] 2002, p. 425.

2 Chase, [1821] 2002.

3 Reynolds, [1839] 2002.

4 Ellis, 2003, p. 243.

5 Ibid., pp. 248–49.

6 Ibid., p. 265.

7 Ibid., p. 251.

8 Williams, 2009.

9 Ellis, 2003, p. 251.

10 "Europe," 2009.

11 Sukumar, 2003, pp. 31–33; Gill et al, 2009.

12 Singer, [1975] 2009, p. 9.

13 Marino et al., 2007, p. 2.

14 Ibid., p. 4.

15 Ibid., p. 5.

16 Ferris, 2009.

17 David, 2009.

18 The debate about what they would look like (Shermer, 2009) may be re-

lated to what they would be like. If they were to look reptilian, perhaps they would also be reptilian.

19 Payne, 1998.

20 Ibid., pp. 94–95.

Bibliography

Abrams, Cooper P., III. 2006. "Polygamy, the Old and New Testaments." *All Experts*, January 19. http://en.allexperts.com/q/Baptists-954/Polygamy-Old-New-Testaments-1.htm.

Alexander, Shana. 2000. *The Astonishing Elephant*. New York: Random House.

Angier, Natalie. 2008. "A Highly Evolved Propensity for Deceit." *New York Times*, Science section, December 22.

Arcadi, Adam Clark, and Richard W. Wrangham. 1999. "Infanticide in Chimpanzees: Review of Cases and a New Within-Group Observation from the Kanyawara Study Group in Kibale National Park." *Primates* 40 (2): 337–51.

Arens, W. 1979. *The Man-eating Myth: Anthropology and Anthropophagy*. New York: Oxford University Press.

Arnhart, Larry. 2008. "Ross, Lincoln, and the Biblical Morality of Slavery." *Darwinian Conservatism*, August 15. http://darwinianconservatism.blogspot.com/2008;-8/ross-lincoln-and-biblical-morality-of.html.

Axelrod, Robert, and William D. Hamilton. 1981. "The Evolution of Cooperation." *Science* 211 (March 27): 1390–96.

Bagemihl, Bruce. 1999. *Biological Exuberance: Animal Homosexuality and Natural Diversity*. New York: St. Martin's Press.

Balcombe, Jonathan P., Neal D. Barnard, and Chad Sandusky. 2004. "Laboratory Routines Cause Animal Stress." *Contemporary Topics, American Association for Laboratory Science* 43:42–51.

Barrett, L., et al. 1999. "Market Forces Predict Grooming Reciprocity in Female Baboons." *Proceedings: Biological Sciences* 266 (April 7):665–70.

Bates, Lucy A., et al. 2008. "Do Elephants Show Empathy?" *Journal of Consciousness Studies* 15 (10, 11): 204–25.

Baur, Bruno. 1992. "Cannibalism in Gastropods." In *Cannibalism: Ecology and Evolution among Diverse Taxa*, ed. Mark A. Elgar and Bernard J. Crespi, 102–27. New York: Oxford University Press.

Beatram, B. C. R. 1975. "Social Factors Influencing Reproduction in Wild Lions." *Journal of Zoology* 177:463–82.

Beaupré, M. G., and U. Hess. 2003. "In my mind, we all smile: A case of in-group favoritism." *Journal of Experimental Social Psychology* 39:371–77.

Bekoff, Marc. 2010. *The Animal Manifesto: Six Reasons for Expanding Our Compassion Footprint*. Novato, CA: New World Library.

Bekoff, Marc, and Jessica Pierce. 2009. *Wild Justice: The Moral Lives of Animals*. Chicago: University of Chicago Press.

Belyaev, Dmitri K. 1979. "Destabilizing selection as a factor in domestication." *Journal of Heredity* 70:301–8.

Biben, Maxeen. 1998. "Squirrel monkey play fighting: Making the case for a cognitive training function for play." In *Animal Play: Evolutionary, Comparative, and Ecological Perspectives*, ed. Marc Bekoff and John A. Byers, 161–82. Cambridge: Cambridge University Press.

Blakeslee, Sandra. 2006. "Cells That Read Minds." *New York Times*, January 10.

Boehm, Christopher. 1999. *Hierarchy in the Forest: The Evolution of Egalitarian Behavior*. Cambridge, MA: Harvard University Press.

———. 2008. "Purposive Social Selection and the Evolution of Human Altruism." *Cross-Cultural Research* 42 (November):319–52.

Boesch, Christophe. 2003. "Complex Cooperation Among Taï Chimpanzees." In *Animal Social Complexity: Intelligence, Culture, and Individualized Societies*, ed. Frans B. M. de Waal and Peter L. Tyack, 93–110. Cambridge, MA: Harvard University Press.

———. 1994. "Cooperative Hunting in Wild Chimpanzees." *Animal Behaviour* 48:653–67.

Boesch, Christophe, and Hedwige Boesch-Achermann. 2000. *The Chimpanzees of the Taï Forest: Behavioural Ecology and Evolution*. Oxford: Oxford University Press.

———. 1981. "Sex Differences in the Use of Natural Hammers by Wild Chimpanzees: A Preliminary Report." *Journal of Human Evolution* 10:585–93.

Bok, Sissela. 1978. *Lying: Moral Choice in Public and Private Life*. New York: Pantheon.

Bond, Charles F., and Michael Robinson. 1988. "The Evolution of Deception." *Journal of Nonverbal Behavior* 12 (December): 295–307.

Boucher, David, and Paul Kelly, eds. 1994. *The Social Contract from Hobbes to Rawls*. New York: Routledge.

Bourne, Geoffrey H. 1974. *Primate Odyssey.* New York: G. P. Putnam's Sons.

Bradshaw, G. A., et al. 2005. "Elephant Breakdown." *Nature* 433 (February 24): 807.

Bräuer, J., et al. 2006. "Making Inference About the Location of Hidden Foods: Social Dog, Causal Ape." *Journal of Comparative Psychology* 120:38–47.

Breed, Michael D. 2001. "Deceit Versus Honest Signalling." http://www.animalbehavioronline.com/deceit.html.

Breman, Joel G., et al. 1999. "A Search for Ebola Virus in Animals in the Democratic Republic of Congo and Cameroon: Ecologic, Virologic, and Serologic Surveys, 1997–1980." *Journal of Infectious Diseases* 179 (suppl. I): S139–S147.

Bronner, Ethan. 2009. "The Bullets in My In-box." *New York Times,* Week in Review, January 25, 1, 4.

Brosnan, Sarah F., Hillary C. Schiff, and Frans B. M. de Waal. 2005. "Tolerance for Inequity May Increase with Social Closeness in Chimpanzees." *Proceedings: Biological Sciences* 272 (1560): 253–58.

Brosnan, Sarah F., and Frans B. M. de Waal. 2003. "Monkeys Reject Unequal Pay." *Nature* 425:297–99.

Bull, J. J., and E. L. Charnov. 1985. "On Irreversible Evolution." *Evolution* 39:1149–55.

Burghardt, Gordon M. 1998. "The Evolutionary Origins of Play Revisited: Lessons from Turtles." In *Animal Play: Evolutionary, Comparative, and Ecological Perspectives,* ed. Marc Bekoff and John A. Byers, 1–26. Cambridge: Cambridge University Press.

———. 2005. *The Genesis of Animal Play: Testing the Limits.* Cambridge, MA: MIT Press.

Butinsky, Tom. 2000. "Africa's Endangered Great Apes." *African Environment & Wildlife* 8 (June): 33–42.

Byrne, Richard. 2005. "Animal Evolution: Foxy Friends." *Current Biology,* February 8: R86, R87.

Byrne, Richard W., and Andrew Whiten. 1987. "The Thinking Primate's Guide to Deception." *New Scientist,* December 3: 54–57.

Carey, Benedict. 2006. "Message from Mouse to Mouse: I Feel Your Pain." *New York Times,* July 4.

Caufield, Catherine. 1985. *In the Rainforest.* New York: Alfred A. Knopf.

Chadwick, Douglas H. 1992. *The Fate of the Elephant.* San Francisco: Sierra Club Books.

Chase, Owen. [1821] 2002. "The Essex Wrecked by a Whale." In *Moby-Dick,*

Norton Critical 2nd ed., ed. Hershel Parker and Harrison Hayford, 565–70. New York: W. W. Norton.

Cheney, Dorothy L., and Robert M. Seyfarth. 2007. *Baboon Metaphysics: The Evolution of a Social Mind.* Chicago: University of Chicago Press.

———. 1990. *How Monkeys See the World: Inside the Mind of Another Species.* Chicago: University of Chicago Press.

Church, Russell M. 1959. "Emotional Reactions of Rats to the Pain of Others." *Journal of Comparative and Physiological Psychology* 52:132–34.

Cline, Austin. 2006. "Thomas Jefferson, Slavery, and Morality." *Austin's Atheism Blog,* June 4. http://atheism.about.com/b/2006/06/04/thomas-jefferson-slavery-and-morality.htm.

Coimbra-Filho, Adelmar F., and Russell A. Mittermeier. 1977. "Conservation of the Brazilian Lion Tamarins (*Leontopithecus rosalia*)." In *Primate Conservation,* ed. Prince Rainier III and Geoffrey H. Bourne, 59–91. New York: Academic Press.

Committee on Selected Biological Problems in the Humid Tropics. 1982. *Ecological Aspects of Development in the Humid Tropics.* Washington, D.C.: National Academy Press.

Connolly, Kevin, and Margaret Martlew, eds. 1999. *Psychologically Speaking: A Book of Quotations.* Leicester: BPS Books.

Cook, M., and S. Mineka. 1989. "Observational conditioning of fear to fear-relevant versus fear-irrelevant stimuli in rhesus monkeys." *Journal of Abnormal Psychology* 98:448–459.

Coppinger, Lorna, and Raymond Coppinger. 2004. *Dogs: A Startling New Understanding of Canine Origin, Behavior, and Evolution.* London: Crosskeys Select.

Cudd, Ann. 2007. "Contractarianism." *Stanford Encyclopedia of Philosophy.* http://plato.stanford.edu/entries/contractarianism/.

Curry, Oliver. 2006. "Who's Afraid of the Naturalistic Fallacy?" *Evolutionary Psychology* 4:234–247.

Da Fonseca, Gustavo A. B. 1985. "Observations on the Ecology of the Muriqui (*Brachyteles arachnoides* E. Geoffroy 1806): Implications for its Conservation." *Primate Conservation* 5 (January): 48–52.

Dally, Joanna M. 2007. "Don't Call Me Birdbrained." *New Scientist* (June 23): 34–37.

Dally, Joanna M., Nathan J. Emery, and Nicola S. Clayton. 2006. "Food-Caching Western Scrub-Jays Keep Track of Who Was Watching When." *Science* 312 (June 16): 1662–1665.

Daly, Martin, and Margo Wilson. 1988. *Homicide.* New York: Aldine de Gruyter.

Darwin, Charles. 1871. *The Descent of Man and Selection in Relation to Sex.* London: John Murray.

———. [1872] 1998. *The Expression of the Emotions in Man and Animals.* Ed. Paul Ekman. Oxford: Oxford University Press.

———. [1859] 1996. *On the Origin of Species.* Ed. Gillian Beer. Oxford: Oxford University Press.

David, Ariel. 2009. "Vatican Looks to Heavens for Signs of Alien Life." Associated Press, November 10. http://enews.earthlink.net/article.

Dawkins, Richard. [1976] 2006. *The Selfish Gene.* Oxford: Oxford University Press.

De Assumpcao, C. Torres. 1983. "Ecological and Behavioural Information on *Brachyteles arachnoides.*" *Primates* 24 (October): 584–93.

Decety, Jean, et al. 2004. "The Neural Bases of Cooperation and Competition: An fMRI Investigation." *NeuroImage* 23:744–75.

de Montellano, B. R. O. 1978. "Aztec Cannibalism: An Economic Necessity?" *Science* 200:611–17.

Dennett, Daniel C. 1988. "The Intentional Stance in Theory and Practice." In *Machiavellian Intelligence: Social Expertise and the Evolution of Intellect in Monkeys, Apes, and Humans,* ed. Richard Byrne and Andrew Whiten, 180–202. Oxford: Clarendon Press

———. 1983. "Intentional Systems in Cognitive Ethology: The 'Panglossian Paradigm' Defended." *Behavioral and Brain Sciences* 6:343–90.

DePaulo, Bella A. 2004. "The Many Faces of Lies." In *The Social Psychology of Good and Evil,* ed. A. G. Miller, 303–26. New York: Pantheon.

Descartes, René. [1637] 1993. "Discourse on Method." In *Environmental Ethics: Divergence and Convergence,* ed. S. J. Armstrong and R. G. Botzler, 281–85. New York: McGraw-Hill.

Dietz, Lou Ann. 1984. "Gold Lion Tamarins Reintroduced to Wild!" *Focus* 6 (July/August): 6.

Douglas-Hamilton, Iain, and Oria Douglas-Hamilton. 1975. *Among the Elephants.* New York: Viking Press.

Drea, Christine M., and Laurence G. Frank. 2003. "The Social Complexity of Spotted Hyenas." In *Animal Social Complexity: Intelligence, Culture, and Individualized Societies,* ed. Frans B. M. de Waal and Peter L. Tyack, 121–48. Cambridge, MA: Harvard University Press.

Dugatin, Lee Alan, ed. 2006. *The Altruism Equation: Seven Scientists Search for the Origins of Goodness.* Princeton, NJ: Princeton University Press.

———. 1999. *Cheating Monkeys and Citizen Bees: The Nature of Cooperation in Animals and Humans.* New York: Free Press.

Ekman, Paul. [1985] 2009. *Telling Lies.* New York: W. W. Norton.

Elgar, Mark A. 1992. "Sexual Cannibalism in Spiders and Other Invertebrates." In *Cannibalism: Ecology and Evolution Among Diverse Taxa*, ed. Mark A. Elgar and Bernard J. Crespi, 128–55. New York: Oxford University Press.

Ellis, Richard. 2003. *The Empty Ocean: Plundering the World's Marine Life.* Washington: Island Press / Shearwater Books.

Emery, Nathan J., and Nicola S. Clayton. 2001. "Effects of Experience and Social Context on Prospective Caching Strategies by Scrub Jays." *Nature* 414 (November 22): 443–46.

Enard, Wolfgang, et al. 2002. "Molecular evolution of FOXP2, a gene involved in speech and language." *Nature* 41:869–72.

Epley, N., and E. Whitchurch. 2008. "Mirror, mirror on the wall: Enhancement in self-recognition." *Personality and Social Psychology Bulletin* 34:1159–70.

"Europe: Whale Enemy No. 1." 2009. *PR Newswire*, June 9.

Evans, E. P. n.d. *The Criminal Prosecution and Capital Punishment of Animals.* London: Faber and Faber.

Fagen, R. 1981. *Animal Play Behavior.* Oxford: Oxford University Press.

Ferris, Timothy. 2009. "Worlds Apart: Seeking New Earths." *National Geographic*, December, 91–93.

Finkelman, Paul. 2001. *Slavery and the Founders: Race and Liberty in the Age of Jefferson.* Armonk, NY: M. E. Sharpe.

Fisher, Helen. 1992. *Anatomy of Love: The Natural History of Monogamy, Adultery, and Divorce.* New York: W. W. Norton.

———. 2004. *Why We Love: The Nature and Chemistry of Romantic Love.* New York: Henry Holt and Company.

Fisher, R. A. 1930. *The Genetical Theory of Natural Selection.* Oxford: Clarendon Press.

Flombaum, Jonathan L., and Laurie R. Santos. 2005. "Rhesus Monkeys Attribute Perceptions to Others." *Current Biology* 15 (March 8): 447–52.

Fossey, Dian. 1983. *Gorillas in the Mist.* Boston: Houghton Mifflin.

Fouts, Roger, and Stephen Turkel Mills. 1997. *Next of Kin: What Chimpanzees Have Taught Me About Who We Are.* New York: William Morrow.

Frank, Laurence G. 1986. "Social Organization of the Spotted Hyaena (*Crocuta crocuta*). I. Demography." *Animal Behaviour* 35:1500–1509.

Frank, S. A. 2005. "Social Selection." In *Evolutionary Genetics: Concepts and Case Studies*, ed. C. W. Fox and J. B. Wolf, 350–63. Oxford: Oxford University Press.

Freud, Sigmund. [1925] 1953. "Some Psychical Consequences of the Anatomical

Distinctions Between the Sexes." In *The Standard Edition of the Complete Psychological Works of Sigmund Freud*, ed. James Strachey. London: Hogarth Press.

Furuichi, Takeshi. 1989. "Social Interactions and the Life History of Female *Pan paniscus* in Wamba, Zaïre." *International Journal of Primatology* 10:173–97.

Gale, U Toke. 1974. *Burmese Timber Elephant.* Rangoon, Burma: Trade.

Ganguli, Ishani. 2009. "Mice Show Evidence of Empathy." *Scientist*, June 30. http://www.the-scientist.com/news/print/23764/.

Gao, Feng, et al. 1999. "Origin of HIV-1 in the Chimpanzee *Pan troglodytes troglodytes.*" *Nature* 397 (February): 436–41.

Gardner, R. Allen, and Beatrix T. Gardner, eds. 1989. *Teaching Sign Language to Chimpanzees.* Albany, NY: SUNY Press.

Gardner, Simon. 2008. "Whale Is 'Just Another Animal': Iceland Decries 'Principle of Survival of the Cutest.'" *Gazette*, June 26.

Gauthier, David. 1986. *Morals by Agreement.* Oxford: Oxford University Press.

Georges, Alain-Jean. 1999. "Ebola Hemorrhagic Fever Outbreaks in Gabon, 1995–1997: Epidemiologic and Health Control Issues." *Journal of Infections Diseases* 179 (suppl. I): S65–S75.

Gill, Jacquelyn L., et al. 2009. "Pleistocene Megafaunal Collapse, Novel Plant Communities, and Enhanced Fire Regimes in North America." *Science* 326 (November 20): 1100–1103.

Gilligan, Carol. [1982] 1993. *In a Different Voice.* Cambridge, MA: Harvard University Press.

Goodall, Jane. 1968. "Behaviour of Free-Living Chimpanzees of the Gombe Stream Area." *Animal Behaviour Monographs* 1:163–311.

———. 2001. *Beyond Innocence: An Autobiography in Letters, the Later Years.* Ed. Dale Peterson. Boston: Houghton Mifflin.

———. 1986. *The Chimpanzees of Gombe: Patterns of Behavior.* Cambridge, MA: Harvard University Press.

Goodall, Jane, and Phillip Berman. 1999. *Reason for Hope: A Spiritual Journey.* New York: Warner Books.

Grandin, Temple, and Catherine Johnson. 2005. *Animals in Translation.* New York: Simon and Schuster.

———. 2009. *Animals Make Us Human: Creating the Best Life for Animals.* Boston: Houghton Mifflin.

Green, Steven, and Karen Minkowski. 1977. "The Lion-tailed Monkey and Its South Indian Rain Forest Habitat." In *Primate Conservation*, ed. Prince Rainier III and Geoffrey Bourne, 289–337. New York: Academic Press.

Griffin, Donald R. [1976] 1981. *The Question of Animal Awareness: Evolutionary Continuity of Mental Experience.* New York: Rockefeller University Press.

Gyger, M., and P. Marler. 1988. "Food calling in the domestic fowl, *Gallus gallus*, the role of external referents and deception." *Animal Behaviour* 36:358–65.

Haan, Norma. 1975. "Hypothetical and Actual Moral Reasoning." *Journal of Personality and Social Psychology* 32:255–70.

Hahn, Beatrice, et al. 2000. "AIDS as a Zoonosis: Scientific and Public Health Implications." *Science* 287 (January 28): 607–14.

Haldane, J. B. S. 1955. "Population Genetics." *New Biology* 18:34–51.

Hamilton, W. D. 1963. "The Evolution of Altruistic Behavior." *American Naturalist* 97:354–56.

———. 1964. "The Genetical Evolution of Social Behavior." *Journal of Theoretical Biology* 7 (1): 1–52.

Harden, Blaine. 2008. "A Clash of Views of Whale-Loving: Creature Is a Delicacy in Japan, a Cause in the West." *Washington Post*, January 26.

Hare, Brian, et al. 2005. "Social Cognitive Evolution in Captive Foxes Is a Correlated By-product of Experimental Domestication." *Current Biology*, February 8, 226–30.

Harner, M. 1977. "The Ecological Basic for Aztec Sacrifice." *American Ethnologist* 4:117–35.

Harris, Marvin. 1989. *Our Kind: Who We Are, Where We Came From, Where We Are Going.* New York: HarperCollins.

Hart, Benjamin L., and Lynette A. Hart. 1992. "Reciprocal Allogrooming in Impala, *Aepyceros melampus*." *Animal Behaviour* 44:1073–83.

Hauser, Marc D. 2006. *Moral Minds: How Nature Designed Our Universal Sense of Right and Wrong.* New York: HarperCollins.

"Hayden Panettiere, Dr. Roger Payne, and Others Call on the International Whaling Commission to End All Commercial and Scientific Whaling." 2009. *PR Newswire*, June 22.

Heinrich, Berndt, and Rachel Smokler. 1998. "Play in Common Ravens (*Corvus corax*)." In *Animal Play: Evolutionary, Comparative, and Ecological Perspectives*, ed. Marc Bekoff and John A. Byers, 27–44. Cambridge: Cambridge University Press.

Heinsohn, Robert, and Craig Packer. 1995. "Complex Cooperative Strategies in Group-Territorial African Lions." *Science*, September 1, 1260–63.

Helo, Ari, and Peter Onuf. 2003. "Jefferson, Morality, and the Problem of Slavery." *William and Mary Quarterly* 60 (July): 583–614.

Hiraiwa-Hasegawa, Mariko. 1999. "Cannibalism Among Non-Human Primates." In *Cannibalism: Ecology and Evolution Among Diverse Taxa*, ed. Mark A. Elgar and Bernard J. Crespi, 323–38. New York: Oxford University Press.

Hodgson, G., and T. Knudsen. 2006. "The Nature and Units of Social Selection." *Journal of Evolutionary Economics* 16:477–89.

Hoffman, Martin L. 1981. "Is Altruism Part of Human Nature?" *Journal of Personality and Social Psychology* 40 (1): 121–37.

Hohmann, Gottfried, and Barbara Fruth. 1993. "Field Observations on Meat Sharing Among Bonobos (*Pan paniscus*)." *Folia Primatologica* 60:225–29.

Hölldobler, Bert, and E. O. Wilson. 2009. *The Superorganism: The Beauty, Elegance, and Strangeness of Insect Societies*. New York: W. W. Norton.

Holstein, Constance. 1976. "Development of Moral Judgment: A Longitudinal Study of Males and Females." *Child Development* 47:51–61.

Horowitz, Alexandra. 2009. *Inside of a Dog: What Dogs See, Smell, and Know*. New York: Scribner.

Hrdy, Sarah Blaffer. 2009. *Mothers and Others: The Evolutionary Origins of Mutual Understanding*. Cambridge, MA: Harvard University Press.

Huxley, Thomas Henry. [1894] 2009. *Evolution and Ethics*. Ed. Michael Ruse. Princeton, NJ: Princeton University Press.

Idani, Gen'ichi. 1991a. "Cases of Inter-Unit Group Encounters in Pygmy Chimpanzees at Wamba, Zaïre." In *Primatology Today: Proceedings of the XIIIth Congress of the International Primatological Society*, ed. Akiyoshi Ehara et al. 235–38. Amsterdam: Elsevier.

———. 1991b. "Social Relationships Between Immigrant and Resident Bonobo (*Pan paniscus*) Females at Wamba, Zaïre." *Folia Primatologica* 57:83–95.

Johns, Andrew. 1985. "Currrent Status of the Southern Bearded Saki (*Chiropotes satanas satanas*)." *Primate Conservation* 5 (January): 28.

Johns, Andrew, and J. M. Ayres. 1987. "Southern Bearded Sakis Beyond the Brink." *Oryx* 21 (July): 164–67.

Johnson, Dominic D. P., Pavel Stopka, and David W. Macdonald. 2003. "Ideal Flea Constraints on Group Living: Unwanted Public Goods and the Emergence of Cooperation." *Behavioral Ecology* 15 (1): 181–86.

Jolly, Alison. 1987. "Madagascar: A World Apart." *National Geographic* 171 (February): 149–83.

Judson, Olivia. 2008. "Wanted: Intelligent Aliens, for a Research Project." *New York Times*, October 1.

Kaminski, J., J. Call, and J. Fischer. 2004. "Word learning in a domestic dog: Evidence for 'fast mapping.'" *Science* 304:1682–83.

Kano, Takayoshi. 1992. *The Last Ape: Pygmy Chimpanzee Behavior and Ecology*. Trans. Evelyn Ono Vineberg. Stanford, CA: Stanford University Press.

———. 1979. "A Pilot Study on the Ecology of Pygmy Chimpanzees *Pan paniscus*." In *The Great Apes*, ed. D. A. Hamburg and E. R. McCrown, 23–36. Menlo Park, CA: Benjamin Cummings.

Kaplan, Matt. 2006. "Make Love, Not War." *New Scientist*, December 2, 40–43.

Karanth, K. Ullas. 1983. "Ecological Status of the Lion-tailed Macaque and Its Rainforest Habitats in Karnataka, India." *Primate Conservation* 6 (July): 73–84.

Karr, James R. 1973. "Ecological and Behavioural Notes on the Liontailed Macaque (*Macaca silenus*) in South India." *Journal of the Bombay Natural History Society* 70 (April): 191–92.

Kavanagh, Michael. 1983. *A Complete Guide to Monkeys, Apes and Other Primates.* New York: Viking Press.

———. 1972. "Food-Sharing Behavior Within a Group of Douc Monkeys (*Pygathrix nemaeus nemaeus*)." *Nature* 239 (October): 406–7.

Kistler, John M. 2006. *War Elephants.* Westport, CT: Praeger.

Klein, R. G. 1999. *The Human Career: Human Biological and Cultural Origins.* Chicago: University of Chicago Press.

Kluger, Jeffrey, and Coco Masters. 2006. "How to Spot a Liar." *Time*, August 28, 46–48.

Kosfeld, Michael, et al. 2005. "Oxytocin Increases Trust in Humans." *Nature* 435 (June 2): 673–77.

Krueger, J. I. 2007. "From social projection to social behavior." *European Review of Social Psychology* 18:1–35.

Kruuk, Hans. 1972. *The Spotted Hyena: A Study of Predation and Social Behavior.* Chicago: University of Chicago Press.

Kummer, Hans. 1995. *In Quest of the Sacred Baboon: A Scientist's Journey.* Trans. Ann M. Biederman-Thorson. Princeton, NJ: Princeton University Press.

———. 1968. *Social Organization of the Hamadryas Baboons: A Field Study.* Chicago: University of Chicago Press.

Kunz, T. H., et al. 1994. "Allomaternal Care: Helper-Assisted Birth in the Rodriguez Fruit Bat, *Pteropus rodricensis* (Chiroptera: Pteropodidae)." *Journal of Zoology* 232:691–700.

Langford, Dale J., et al. 2006. "Social Modulation of Pain as Evidence for Empathy in Mice." *Science* 312 (June 30): 1967–70.

Lee, Richard B. 1979. *The !Kung San: Men, Women, and Work in a Foraging Society.* Cambridge: Cambridge University Press.

———. 1982. "Politics, Sexual and Non-Sexual, in an Egalitarian Society." In *Politics and History in Band Societies*, ed. Eleanor Leacock and Richard Lee, 37–59. Cambridge: Cambridge University Press.

Leiber, Justin. 2006. "Instinctive Incest Avoidance: A Paradigm Case for Evolutionary Psychology Evaporates." *Journal for the Theory of Social Behaviour* 36 (4): 369–88.

Lerner, Gerda. 1986. *The Creation of Patriarchy.* Oxford: Oxford University Press.

Lewin, Roger. 1987. "Do Animals Read Minds?" *Science* 238:1350–51.

Lewin, Roger, and Robert A. Foley. 2004. *Principles of Human Evolution.* New York: Wiley-Blackwell.

Lewis, C. S. 1964. *The Discarded Image: An Introduction to Medieval and Renaissance Literature.* Cambridge: Cambridge University Press.

Lieberman, Philip. 2007. "The Evolution of Human Speech, Its Anatomical and Neural Bases." *Current Anthropology* 48 (February): 39–53.

Lippold, Lois K. 1977. "The Douc Langur: A Time for Conservation." In *Primate Conservation*, ed. Prince Rainier III and Geoffrey H. Bourne, 513–38. New York: Academic Press.

Lloyd, James E. 1965. "Aggressive Mimicry in Photuris: Firefly Femmes Fatales." *Science* 149 (August 6): 653–54.

Lorenz, Konrad. [1963] 1966. *On Aggression.* Trans. Marjorie Kerr Wilson. New York: Harcourt Brace & World.

Lovejoy, Thomas E. 1980. "Discontinuous Wilderness: Minimum Areas for Conservation." *Parks* 5:13–15.

Lyons, Derek E., Laurie E. Santos, and Frank C. Keil. 2006. "Reflections of Other Minds: How Primate Social Cognition Can Inform the Function of Mirror Neurons." *Current Opinion in Neurobiology* 16:230–34.

MacAndrew, Alec. 2009. "FOXP2 and the Evolution of Language." http://www.evolutionpages.com/FOXP2_language.

MacFarquhar, Neil. 2008. "In a First, Gay Rights Are Pressed at the U.N." *New York Times*, December 19, A13.

MacLean, Paul. 1990. *The Triune Brain in Evolution: Role in Paleocerebral Functions.* New York: Plenum Press.

Mallison, Jeremy. 1984. "Lion Tamarins' Survival Hangs in Balance." *Oryx* 18 (April): 72–78.

Marino, Lori, et al. 2007. "Cetaceans Have Complex Brains for Complex Cognition." *PLoS Biology* 5 (5): e139.

Masserman, Jules H., Stanley Wechkin, and William Terris. 1964. " 'Altruistic' Behavior in Rhesus Monkeys." *American Journal of Psychiatry* 121:584–85.

McGlone, John. 2002. *Pig Production: Biological Principles and Application.* Clifton Park, NY: Delmar Learning.

Mead, Margaret. 1949. *Male and Female: A Study of the Sexes in a Changing World.* New York: William Morrow.

Mech, L. David. 1995. "Alpha Status, Dominance, and Division of Labor in Wolf Packs." *Canadian Journal of Zoology* 77:1192–1203.

———. 2000. "Leadership in Wolf, *Canis lupus*, Packs." *Canadian Field-Naturalist* 114 (2): 259–63.

Melville, Herman. [1851] 2002. *Moby-Dick.* Norton Critical 2nd ed. Ed. Hershel Parker and Harrison Hayford. New York: W. W. Norton.

Mesnick, Sarah L., et al. 2003. "Sperm Whale Social Structure: Why It Takes a Village to Raise a Child." In *Animal Social Complexity: Intelligence, Culture, and Individualized Societies*, ed. Frans B. M. de Waal and Peter L. Tyack, 170–74. Cambridge MA: Harvard University Press.

Miles, H. Lyn White. 1993. "Language and the Orang-utan: The 'Old Person' of the Forest." In *The Great Ape Project: Equality Beyond Humanity*, ed. Paola Cavalieri and Peter Singer. New York: St. Martin's Press.

Millan, Cesar, and Melissa Jo Peltier. 2006. *Cesar's Way: The Natural, Everyday Guide to Understanding and Correcting Common Dog Problems.* New York: Three Rivers Press.

Miller, Gerrit S. 1903. "Seventy New Malasian Mammals." *Smithsonian Miscellaneous Collections* 45 (November), no. 1420.

Miller, John Chester. 1977. *The Wolf by the Ears: Thomas Jefferson and Slavery.* New York: Free Press.

Miller, Michelle N., and John A. Byers. 1998. "Sparring as play in young pronghorn males." In *Animal Play: Evolutionary, Comparative, and Ecological Perspectives*, ed. Marc Bekoff and John A. Byers, 141–60. Cambridge: Cambridge University Press.

Mineka, S., and M. Cook. 1986. "Immunization against the observational conditioning of snake fear in rhesus monkeys." *Journal of Abnormal Psychology* 95:307–18.

Mittermeier, Russell A. 1987. "Monkey in Peril." *National Geographic* 171 (March): 387–95.

———. 1982. "The World's Endangered Primates: An Introduction and a Case Study—the Monkeys of Brazil's Atlantic Forests." In *Primates and the Tropical Rainforest*, ed. Russell A. Mittermeier and Mark J. Plotkin, 11–22. Pasadena: California Institute of Technology.

Mittermeier, Russell A., et al. 1982. "Conservation of Primates in the Atlantic Forest Region of Eastern Brazil." In *1982 International Zoo Yearbook*, vol. 22, ed. P. J. S. Olney, 2–17. London: Zoological Society of London.

Mock, Douglas. 1992. "Dining Respectably." Science 258 (December 18): 1969–70.

Moffett, Mark W. 2010. *Adventures Among Ants: A Global Safari with a Cast of Trillions.* Berkeley: University of California Press.

Moll, Jorge, Richardo de Oliveira-Souza, and Roland Zahn. 2008. "The Neural Basis of Moral Cognition: Sentiments, Concepts, and Values." *Annals of the New York Academy of Sciences* 1124:161–80.

Moore, George Edward. 1903. *Principia Ethica.* Cambridge: Cambridge University Press.

Moss, Cynthia. 1988. *Elephant Memories: Thirteen Years in the Life of an Elephant Family.* Chicago: University of Chicago Press.

Muehlenberg, Bill. 2007. "The Bible, Slavery, and Morality." *CultureWatch*, July 25. http://www.billmuehlenberg.com/2007/07/25/the-bible-slavery-andmorality/.

Murray, Carson M., Emily Wroblewski, and Anne E. Pusey. 2007. "New Case of Intergroup Infanticide in the Chimpanzees of Gombe National Park." *International Journal of Primatology* 28 (February): 23–37.

Myers, Norman. 1984. *The Primary Source: Tropical Forests and Our Future.* New York: W. W. Norton.

Nagel, Thomas. 1974. "What Is It Like to Be a Bat?" *Philosophical Review* 83 (October): 435–50.

Najafi, Sina. 2007. "The Language of the Bees: An Interview with Hugh Raffles." *Cabinet* 25:87+.

Narveson, Jan. 1988. *The Libertarian Idea.* Philadelphia: Temple University Press.

Nisbet, Robert A., Charles H. Page, and Robert Perrin. 1977. *The Social Bond.* 2nd ed. New York: Knopf.

Norrie, Justin. 2007. "Japan Defends Its Whale Slaughter." *Age* (International), November 24, 15.

Nowak, Martin A., et al. 2006. "Five Rules for the Evolution of Cooperation." *Science* 314 (December 8): 1560–62.

O'Connell, Caitlin. 2007. *The Elephant's Secret Sense: The Hidden Life of the Wild Herds of Africa.* New York: Free Press.

O'Connell, Sanjida M. 1995. "Empathy in Chimpanzees: Evidence for Theory of Mind?" *Primates* 36:397–410.

Otter, Ken, and Lorene Ratcliffe. 1996. "Female Initiated Divorce in a Monogamous Songbird: Abandoning Mates for Males of Higher Quality." *Proceedings of the Royal Society of London* B 263 (1368): 351–54.

Owens, Delia, and Mark Owens. 2005. "Comeback Kids." *Natural History*, July/August, 22–25.

Packer, Craig D. 1988. "Constraints on the Evolution of Reciprocity: Lessons from Cooperative Hunting." *Ethology and Sociobiology* 9:137–47.

Packer, Craig D., et al. 1988. "Reproductive Success of Lions." In *Reproductive Success*, ed. T. H. Clutton-Brock, 363–83. Chicago: University of Chicago Press.

Packer, Craig D., and Anne E. Pusey. 1983. "Adaptations of Female Lions to Infanticide by Incoming Males." *American Naturalist* 121:716–28.

Packer, Craig, D. Scheel, and Anne E. Pusey. 1990. "Why Lions Form Groups: Food Is Not Enough." *American Naturalist* 136:1–9.

Panksepp, Jaak. 2005. "Beyond a Joke: From Animal Laughter to Human Joy?" *Science* 308 (April 1): 62–63.

Panksepp, Jaak, and Jeff Burgdorf. 2003. "'Laughing' Rats and the Evolutionary Antecedents of Human Joy?" *Physiology and Behavior* 79:533–47.

Parish, Amy Randall. 1994. "Sex and Food Control in the 'Uncommon Chimpanzee': How Bonobo Females Overcome a Phylogenetic Legacy of Male Dominance." *Ethology and Sociobiology* 15 (3): 157–79.

Parker, Andrew. 2003. *In the Blink of an Eye: How Vision Sparked the Big Bang of Evolution*. New York: Basic Books.

Patterson, Francine, and Eugene Linden. 1981. *The Education of Koko*. New York: Holt, Rinehart, and Winston.

Patterson, T. 1979. "The Behavior of a Group of Captive Pygmy Chimpanzees (*Pan paniscus*)." *Primates* 20 (3): 341–54.

Pauly, Daniel. 2009. "Aquacalypse Now: The End of Fish." *New Republic*, October 7, 24–27.

Payne, Katy. 1998. *Silent Thunder: In the Presence of Elephants*. New York: Simon and Schuster.

———. 2003. "Sources of Social Complexity in the Three Elephant Species." In *Animal Social Complexity: Intelligence, Culture, and Individualized Societies*, ed. Frans B. M. de Waal and Peter L. Tyack, 57–85. Cambridge, MA: Harvard University Press.

Payne, Katy, William R. Langbauer, and Elizabeth M. Thomas. 1986. "Infrasonic Calls of the Asian Elephant (*Elephas maximus*)." *Behavioural Ecology and Sociobiology* 18:297–301.

Pepperberg, Irene Maxine. 1999. *The Alex Studies: Cognitive and Communicative Abilities of Grey Parrots*. Cambridge, MA: Harvard University Press.

Perry, Susan, with Joseph H. Manson. 2008. *Manipulative Monkeys: The Capuchins of Lomas Barbudal*. Cambridge, MA: Harvard University Press.

Peterson, Dale. 1989. *The Deluge and the Ark: A Journey into Primate Worlds*. Boston: Houghton Mifflin.

———. 2003. *Eating Apes*. Berkeley: University of California Press.

———. 2006. *Jane Goodall: The Woman Who Redefined Man*. Boston: Houghton Mifflin.

Peterson, Dale, and Jane Goodall. 1993. *Visions of Caliban: On Chimpanzees and People*. Boston: Houghton Mifflin.

Pfaff, Donald W. 2007. *The Neuroscience of Fair Play: Why We (Usually) Follow the Golden Rule*. New York: Dana Press.

Piaget, Jean, et al. 1932. *The Moral Judgment of the Child*. London: K. Paul, Trench, Trubner.

Pierce, Jeremy. 2005. "The Morality of Slavery." *Parableman*, February 28. http://parablemania.ektopos.com/archives/2005/02/the_morality_of_.html.

Pinker, Steven. 2008. "The Moral Instinct." *New York Times Magazine*, January 13, 32+.

Pinker, Susan. 2008. *The Sexual Paradox: Men, Women, and the Real Gender Gap*. New York: Scribner.

Plooij, Frans X. 2000. "A Slap in the Face." In *The Smile of a Dolphin: Remarkable Accounts of Animal Emotions*, ed. Marc Bekoff, 88. New York: Discovery Books.

"Polygamy: What the Bible Says." 2006. *Eads Home Ministries*. http://www.eadshome.com/polygamy.htm.

Poole, Joyce. 1996. *Coming of Age with Elephants: A Memoir*. New York: Hyperion.

Porat, D., and N. E. Chadwick-Furman. 2004. "Effects of Anemonefish on Giant Sea Anemones: Expansion Behavior, Growth, and Survival." *Hydrobiologia* 530:513–20.

Power, Margaret. 1991. *The Egalitarians, Human and Chimpanzee: An Anthropological View of Social Organization*. Cambridge: Cambridge University Press.

Preston, Stephanie D., and Frans B. M. de Waal. 2002. "Empathy: Its Ultimate and Proximate Bases." *Behavioral and Brain Sciences* 25:1–72.

Price, Charles Edwin. 1992. *The Day They Hung the Elephant*. Johnson City, TN: The Overmountain Press.

Pronin, E. 2008. "How we see ourselves and how we see others." *Science* 320:1177–80.

Pusey, Anne, et al. 2008. "Severe Aggression Among Female *Pan troglodytes schweinfurthii* at Gombe National Park, Tanzania." *International Journal of Primatology* 29:949–73.

Raby, C. R., et al. 2007. "Planning for the Future by Western Scrub-Jays." *Nature* 445 (February 22): 919–21.

Reynolds, J. N. [1839] 2002. "Mocha Dick: Or the white whale of the Pacific: A leaf from a manuscript journal." In *Moby-Dick*, Norton Critical 2nd ed., ed. Hershel Parker and Harrison Hayford, 549–65. New York: W. W. Norton.

Rice, George E., Jr. 1964. "Aiding Behavior vs. Fear in the Albino Rat." *Psychological Record* 14:165–70.

Rice, George E., Jr., and Priscilla Gainer. 1962. "'Altruism' in the Albino Rat." *Journal of Comparative and Physiological Psychology* 55 (1): 123–25.

Richard, Alison F. 1982. "The World's Endangered Species: A Case Study of the Lemur Fauna of Madagascar." In *Primates and the Tropical Rainforest*, ed. Russell A. Mittermeier and Mark J. Plotkin, 23–30. Pasadena: California Institute of Technology.

Richard, Alison F., and Robert W. Sussman. 1974. "Future of the Malagasy Lemurs: Conservation or Extinction?" In *Lemur Biology*, ed. Ian Tattersall and Robert W. Sussman, 335–50. Pittsburgh: University of Pittsburgh Press.

Rilling, James K., et al. 2002. "A Neural Basis for Social Cooperation." *Neuron* 35 (July 18): 395–405.

Rizzolatti, Giacomo, and Laila Craighero. 2004. "The Mirror-Neuron System." *Annual Review of Neuroscience* 27:169–92.

Robbins, J. M., and J. I. Kreuger. 2005. "Social projection to ingroups and outgroups: A review and meta-analysis." *Personality and Social Psychology Review* 9:32–47.

Ross, Fred A. [1857] 2009. *Slavery Ordained of God.* LaVerne, TN.

Roughgarden, Joan. 2004. *Evolution's Rainbow: Diversity, Gender, and Sexuality in Nature and People.* Berkeley: University of California Press.

Rousseau, Jean-Jacques. [1762] 1968. *The Social Contract.* Trans. Maurice Cranston. Harmondsworth, UK: Penguin.

Rozin, Paul. 1997. "Moralization." In *Morality and Health*, ed. A. Brandt and P. Rozin. New York: Routledge.

———. 1996. "Towards a psychology of food and eating: From motivation to module to model to marker, morality, meaning, and metaphor." *Current Directions in Psychological Science* 5:18–41.

Rudolf, Volker H. W., and Janis Antonovics. 2007. "Disease Transmission by Cannibalism: Rare Event or Common Occurrence?" *Proceedings of the Royal Society* B 274:1205–10.

Rutte, Claudia, and Michael Taborsky. 2009. "Generalized Reciprocity in Rats." *PLoS Biology* 5 (7): e196.

Rylands, Anthony B., and Russell A. Mittermeier. 1982. "Conservation of Primates in Brazilian Amazonia." In *1982 International Zoo Yearbook*, vol. 22, ed. P. J. S. Olney, 17–37. London: Zoological Society of London.

Sacks, Melinda. 2009. "What Makes Elephants Tick? A Stanford Researcher Unlocks Some Behavioral Mysteries." *Stanford*, September/October. http://www.stanfordalumni.org/news/magazine/2009/sepoct/farm/news/elephant.html.

Sagi, Abraham, and Marin L. Hoffman. 1976. "Empathic Distress in the Newborn." *Developmental Psychology* 12 (2): 175–76.

Santos, Laurie R., Aaron G. Nissen, and Jonathan A. Ferrugia. 2006. "Rhesus monkeys, *Macaca mulatta*, know what others can and cannot hear." *Animal Behaviour* 71:1175–81.

Savage-Rumbaugh, Sue, and Roger Lewin. 1994. *Kanzi: The Ape at the Brink of the Human Mind.* New York: John Wiley and Sons.

Savage-Rumbaugh, Sue, and Kelly McDonald. 1988. "Deception and Social Manipulation in Symbol-Using Apes." In *Machiavellian Intelligence: Social Expertise and the Evolution of Intellect in Monkeys, Apes, and Humans*, ed. Richard Byrne and Andrew Whiten, 224–37. Oxford: Clarendon Press.

Savage-Rumbaugh, Sue, and B. J. Wilkerson. 1978. "Socio-Sexual Behavior in *Pan paniscus* and *Pan troglodytes*: A Comparative Study." *Journal of Human Evolution* 7:327–44.

Scigliano, Eric. 2002. *Love, War, and Circuses: The Age-Old Relationship Between Elephants and Humans.* Boston: Houghton Mifflin.

Sharp, Keith. 2001. "Understanding the Law." Vol. 1, no. 9 (November 1). http://www.christistheway.com.

Sherman, Paul W. 1977. "Nepotism and the Evolution of Alarm Calls." *Science* 197 (September 23): 1246–53.

Shermer, Michael. 2009. "Will E.T. Look Like Us?" *Scientific American*, November 36.

Sherrow, Hogan M., and Sylvia J. Amsler. 2007. "New Intercommunity Infanticides by the Chimpanzees of Ngogo, Kibale National Park, Uganda." *International Journal of Primatology* 28 (February): 9–22.

Shoshani, Jehesekel, and J. F. Eisenberg. 1982. "*Elephas maximus*." *Mammalian Species* 182 (June 18): 1–8.

Shostak, Marjorie. 1981. *Nisa: The Life and Words of a !Kung Woman.* Cambridge, MA: Harvard University Press.

Shu, Weiguo, et al. 2005. "Altered ultrasonic vocalization in mice with a disruption in the *Foxp2* gene." *PNAS* 102 (July 5): 9643–48.

Siebert, Charles. 2006. "An Elephant Crackup?" *New York Times Magazine*, October 8, 42+.

Sikes, Sylvia K. 1971. *The Natural History of the African Elephant.* New York: American Elsevier Publishing.

Silk, Joan B., Robert M. Seyfarth, and Dorothy L. Cheney. 1999. "The Structure of Social Relationships Among Female Savanna Baboons in Moremi Reserve, Botswana." *Behaviour* 136:679–703.

Silverberg, James, and J. Patrick Gray. 1992. "Violence and Peacefulness as Behavioral Potentialities of Primates." In *Aggression and Peacefulness in Humans and Other Primates*, ed. James Silverberg and J. Patrick Gray, 1–36. Oxford: Oxford University Press.

Simon, H. 1990. "A mechanism for social selection and successful altruism." *Science* 250:1665–68.

Simon, Noel, and Paul Geroudet. 1970. *Last Survivors: The Natural History of Animals in Danger of Extinction*. New York: World Publishing.

Singer, Peter. [1975] 2009. *Animal Liberation*. New York: HarperCollins.

Skilton, Nyssa. 2009. "Pretenders Bob Up in Lizard Mating Game." *Canberra Times*. http://www.canberratimes.com.au/news/local/new/general/pretenders-bob-up-in-lizard-mating-game/1449648.aspx.

Slater, Kathy V., C. M. Schaffner, and F. Aureli. 2007. "Embraces for Infant Handling of Spider Monkeys: Evidence for a Biological Market?" *Animal Behaviour* 74:455–61.

Slotow, Rob, Dave Balfour, and Owen Howison. 2001. "Killing of Black and White Rhinoceroses by African Elephants in Hluhluwe-Umfolozi Park, South Africa." *Pachyderm* 31 (July–December): 14–20.

Slowtow, Rob, et al. 2000. "Older Bull Elephants Control Young Males." *Nature* 408 (November 23): 425–26.

Slotow, Rob, and G. van Dyk. 2001. "Role of Delinquent Young 'Orphan' Male Elephants in High Mortality of White Rhinoceros in Pilanesberg National Park, South Africa." *Koedoe* 44 (1): 85–94.

Smith, David Livingstone. 2007. *The Most Dangerous Animal: Human Nature and the Origins of War.* New York: St. Martin's Press.

Smith, J. M. 1964. "Group Selection and Kin Selection." *Nature* 201 (4924): 1145–47.

Smith, Lewis. 2008. "Review of Primates Finds 303 Species Threatened." *Australian*, August 5. http://www.theaustralian.news.com.au/story/0,25197,24131596-2703,00.html.

Smith, Susan M. 1991. *The Black-capped Chickadee: Behavioral Ecology and Natural History.* Ithaca, NY: Comstock Publishing Associates.

Smuts, Barbara B., and J. M. Watanabe. 1990. "Social Relationships and Ritualized Greetings in Adult Male Baboons (*Papio cynocephalus anubis*)." *International Journal of Primatology* 11:147–72.

Stanford, Craig. 2001. *Significant Others: The Ape-Human Continuum and the Quest for Human Nature.* New York: Basic Books.

Stark, W. 1961. "Natural and Social Selection." In *Darwinism and the Study of Society: A Centenary Symposium*, ed. M. Banton, 49–61. London: Tavistock Publications.

Stevens, Jeffrey R., and Marc D. Hauser. 2004. "Why Be Nice? Psychological Constraints on the Evolution of Cooperation." *Trends in Cognitive Sciences* 8 (February): 60–65.

Strayer, F. F. 1992. "The Development of Agonistic and Affiliative Structures in Preschool Play Groups." In *Aggression and Peacefulness in Humans and Other Primates*, ed. James Silverberg and J. Patrick Gray, 150–71. Oxford: Oxford University Press.

Strier, Karen B. 1992a. "Causes and Consequences of Nonaggression in the Woolly Spider Monkey, or Muriqui (*Brachyteles arachnoides*)." In *Aggression and Peacefulness in Humans and Other Primates*, ed. James Silverberg and J. Patrick Gray, 100–116. Oxford: Oxford University Press.

———. 1992b. *Faces in the Forest: The Endangered Muriqui Monkeys of Brazil*. Oxford: Oxford University Press.

Sugiyama, Yukimaru. 1968. "The Ecology of the Lion-tailed Macaque (*Macaca silenus* [Linnaeus])—a Pilot Study." *Journal of the Bombay Natural History Society* 65 (August): 283–93.

Sukumar, Raman. 2003. *The Living Elephants: Evolutionary Ecology, Behavior, and Conservation*. Oxford: Oxford University Press.

Sundström, A. 2008. "Self-assessment of driving skill—a review from a measurement perspective." *Transportation Research Part F* 11:1–9.

Sussman, Robert W., Paul A. Garber, and Jim M. Cheverud. 2005. "Importance of Cooperation and Affiliation in the Evolution of Primate Sociality." *American Journal of Physical Anthroplogy* 128:84–97.

Sussman, Robert W., and Joshua L. Marshack. 2010. "Are Humans Inherently Killers?" Prepublication MS.

Sussman, Robert W., Alison F. Richard, and Gilbert Ravelojaona. 1985. "Mada gascar: Current Projects and Problems in Conservation." *Primate Conservation* 5 (January): 53–59.

Tattersall, Ian. 1982. *The Primates of Madagascar*. New York: Columbia University Press.

Taylor, Charles E., and Michael T. McGuire. 1988. "Reciprocal Altruism: 15 Years Later." *Ethology and Sociobiology* 9:67–72.

Teleki, Geza. 1973. *The Predatory Behavior of Wild Chimpanzees*. East Brunswick, NJ: Bucknell University Press.

Tenaka, Y. 1996. "Social Selection and the Evolution of Animal Signals." *Evolution* 50:512–23.

Tenaza, Richard R. 1975. "Territory and Monogamy Among Kloss' Gibbons (*Hylobates klossii*) in Siberut Island, Indonesia." *Folia Primatologica* 24:60–80.

Tenaza, Richard R., and W. J. Hamilton III. 1971. "Preliminary Observations of the Mentawai Islands Gibbon, *Hylobates klossii.*" *Folia Primatologica* 15:201–11.

Tenaza, Richard R., and Arthur Mitchell. 1985. "Summary of Primate Conservation Problems in the Mentawai Islands, Indonesia." *Primate Conservation* 6 (July): 36–37.

"Ten Commandments." 2008. Wikipedia. http://en.wikipedia.org/w/index.php?title=Ten_Commandments.

Thierry, Bernard. 2000. "Building Elements of Morality Are Not Elements of Morality." In *Evolutionary Origins of Morality: Cross-Disciplinary Perspectives*, ed. Leonard D. Katz, 60–65. Bowling Green, OH: Imprint Academic.

Thomas, Elizabeth Marshall. [1959] 1968. *The Harmless People.* New York: Knopf.

Thornback, Jane, and Martin Jenkins, eds. 1982. *The IUCN Mammal Red Data Book*, Part I. Gland, Switzerland: IUCN.

Trivers, Robert L. 1971. "The Evolution of Reciprocal Altruism." *Quarterly Review of Biology* 46 (March): 35–57.

"Trouser-Wearing Woman Spared Flogging in Sudan." 2009. September 7. http://enews.earthlink.net/channel/news.

Trut, Ludmilla N. 1999. "Early Canid Domestication: The Farm-Fox Experiment." *American Scientist* 87:160–69.

Tucker, Abigail. 2010. "The Truth About Lions." *Smithsonian*, January, 28–39.

Ungerer, Judy A., et al. 1990. "The Early Development of Empathy: Self-Regulation and Individual Differences in the First Year." *Motivation and Emotion* 14 (2): 93–106.

Vallentyne, Peter, ed. 1991. *Contractarianism and Rational Choice.* Cambridge: Cambridge University Press.

Van Hooff, Jan A. R. A. M., and Signe Preuschoft. 2003. "Laughter and Smiling: The Intertwining of Nature and Culture." In *Animal Social Complexity: Intelligence, Culture, and Individualized Societies*, ed. Frans B. M. de Waal and Peter L. Tyack, 260–92. Cambridge, MA: Harvard University Press.

Van Peenen, P. F. D., R. H. Light, and J. F. Duncan. 1971. "Observation on Mammals of Mt. Sontra, South Vietnam." *Mammalia* 35:126–43.

Vedantam, Shankar. 2006. "How Brain's 'Mirrors' Aid Our Social Understanding." *Washington Post*, September 25.

Villa, P., et al. 1986. "Cannibalism in the Neolithic." *Science* 233:431–37.

de Waal, Frans B. M. 1992. "Aggression as a Well-Integrated Part of Primate Social Relationships: A Critique of the Seville Statement on Violence." In *Aggression and Peacefulness in Humans and Other Primates*, ed. James Silverberg and J. Patrick Gray, 37–56. Oxford: Oxford University Press.

———. 1997. *Bonobo: The Forgotten Ape.* Berkeley: University of California Press.

———. 1982. *Chimpanzee Politics: Power and Sex Among Apes.* New York: Harper and Row.

———. 1988. "The Communication Repertoire of Captive Bonobos (*Pan paniscus*), Compared to That of Chimpanzees." *Behaviour* 106:183–251.

———. 1996. *Good Natured: The Origins of Right and Wrong in Humans and Other Animals.* Cambridge, MA: Harvard University Press.

———. 2005. *Our Inner Ape: A Leading Primatologist Explains Why We Are Who We Are.* New York: Riverhead Books.

Walter, Alex. 2006. "The Anti-Naturalistic Fallacy: Evolutionary Moral Psychology and the Insistence of Brute Facts." *Evolutionary Psychology* 4:33–48.

Warneken, Felix, et al. 2007. "Spontaneous Altruism by Chimpanzees and Young Children." *PLoS Biology* 5 (7): e184.

Watson, Duncan M. 1998. "Kangaroos at play: Play behaviour in the Macropodoidea." In *Animal Play: Evolutionary, Comparative, and Ecological Perspectives*, ed. Marc Bekoff and John A. Byers, 45–96. Cambridge: Cambridge University Press.

Watson, Karli, and Michael L. Platt. 2006. "Fairness and the Neurobiology of Social Cognition: Commentary on 'Nonhuman species' reactions to inequity and their implications for fairness' by Sarah Brosnan." *Social Justice Research* 19 (June): 186–93.

Weeks, Kevin, and Phyllis Karas. 2006. *Brutal: The Untold Story of My Life Inside Whitey Bulger's Irish Mob.* New York: HarperCollins.

Weissengrubber, G. E., et al. 2006. "The Structure of the Cushions in the Feet of African Elephants (*Loxodonta africana*)." *Journal of Anatomy* 209 (December): 781–92.

"Whaling." 2009. Wikipedia (last modified November 19). http://en.wikipedia.org/w/index.php?title=Whaling.

Whitehead, Hal. 2003. "Society and Culture in the Deep and Open Ocean: The Sperm Whale and Other Cetaceans." In *Animal Social Complexity: Intelligence, Culture, and Individualized Societies*, ed. Frans B. M. de Waal and Peter L. Tyack, 444–64. Cambridge, MA: Harvard University Press.

Whiting, Martin J. 2000. "The Augrabies Flat Lizard." In *Augrabies Splendour: A Guide to the Natural History of the Augrabies Falls National Park and the Riemvasmaak Wildlife Area*, ed. P. van der Walt, 56–63. Privately published.

Whiting, Martin J., Jonathan K. Webb, and J. Scott Keogh. 2009. "Flat Lizard Female Mimics Use Sexual Deception in Visual but Not Chemical Signals." *Proceedings of the Royal Society* B 276:1585–91.

Whyte, Martin King. 1978. "Cross-cultural Codes Dealing with the Relative Status of Women." *Ethnology* 17 (April): 211–37.

Wiedenmayer, Christoph. 1997. "Causation of the Ontogenetic Development of Stereotypic Digging in Gerbils." *Animal Behaviour* 53 (3): 461–70.

Wiley, R. H. 1974. "Evolution of social organization and life history patterns among grouse (Aves: Tetraonidae)." *Quarterly Review of Biology* 49 (3): 201–27.

———. 1973. "Territoriality and non-random mating in sage grouse, *Centrocercus urophasianus*." *Animal Behavior Monographs* 6 (2): 85–169.

Wilkinson, Gerald S. 1988. "Reciprocal Altruism in Bats and Other Mammals." *Ethology and Sociobiology* 9:85–100.

———. 1984. "Reciprocal Food Sharing in the Vampire Bat." *Nature* 308 (March 8): 181–84.

Williams, Brian. 2009. "Japanese Harpoons Kill 8,175." *Courier Mail*, June 20, 44.

Wilson, Edward O. [1975] 2000. *Sociobiology: The New Synthesis*. 25th Anniversary ed. Cambridge, MA: Harvard University Press.

Wilson, Michael L., Nicholas F. Britton, and Nigel R. Franks. 2002. "Chimpanzees and the Mathematics of Battle." *Proceedings of the Royal Society of London* 269:1107–12.

Wilson, Michael L., Marc D. Hauser, and Richard W. Wrangham. 2001. "Does Participation in Intergroup Conflict Depend on Numerical Assessment, Range Location, or Rank for Wild Chimpanzees?" *Animal Behaviour* 61:1203–16.

Wilson, Michael L., William R. Wallauer, and Anne E. Pusey. 2004. "New Cases of Intergroup Violence Among Chimpanzees in Gombe National Park, Tanzania." *International Journal of Primatology* 25 (June): 523–49.

Wilson, Michael L., and Richard W. Wrangham. 2003. "Intergroup Relations in Chimpanzees." *Annual Review of Anthropology* 32:363–92.

Wolfheim, Jaclyn. 1983. *Primates of the World: Distribution, Abundance, and Conservation*. Seattle: University of Washington Press.

Wolpoff, M. H. 1999. *Paleoanthropology*. 2nd. ed. Boston: McGraw-Hill.

"Women's Suffrage." 2009. Wikipedia (last modified October 5). http://en.wikipedia.

Woodruff, Guy, and David Premack. 1979. "Intentional Communication in the Chimpanzee: The development of deception." *Cognition* 7:333–62.

Wrangham, Richard W. 1975. "The Behavioral Ecology of Chimpanzees in Gombe National Park, Tanzania." Ph.D. diss., Cambridge University.

———. 2009. *Catching Fire: How Cooking Made Us Human*. New York: Basic Books.

———. 2000. "Making a Baby." In *The Smile of a Dolphin: Remarkable Accounts of Animal Emotions*, ed. Marc Bekoff, 34. New York: Discovery Books.

Wrangham, Richard W., William C. McGrew, and Frans B. M. de Waal. 1994. "The Challenge of Behavioral Diversity." In *Chimpanzee Cultures*, ed. Richard W. Wrangham et al. Cambridge, MA: Harvard University Press.

Wrangham, Richard W., and Dale Peterson. 1996. *Demonic Males: Apes and the Origins of Human Violence*. Boston: Houghton Mifflin.

Wrangham, Richard W., and Emilie van Zinnicq Bergmann Riss. 1990. "Rates of Predation of Mammals by Gombe Chimpanzees, 1972–1975." *Primates* 31 (2): 157–70.

Wrangham, Richard W., and Michael L. Wilson. 2004. "Collective Violence: Comparisons Between Youths and Chimpanzees." *Annals of the New York Academy of Sciences* 1036:233–56.

Wrangham, Richard W., Michael L. Wilson, and Martin N. Muller. 2006. "Comparative Rates of Violence in Chimpanzees and Humans." *Primates* 47:14–16.

Wright, Robert. 1994. *The Moral Animal: Evolutionary Psychology and Everyday Life*. New York: Pantheon.

Zahn-Waxler, Carolyn, et al. 1992. "Development of Concern for Others." *Developmental Psychology* 28 (1): 126–36.

Zahn-Waxler, Carolyn, Sarah L. Friedman, and E. Mark Cummings. 1983. "Children's Emotions and Behaviors in Response to Infants' Cries." *Child Development* 54:1522–28.

Zahn-Waxler, Carolyn, JoAnn L. Robinson, and Robert N. Emde. 1992. "The Development of Empathy in Twins." *Developmental Psychology* 28 (6): 1038–47.

Zuckerman, Wendy. 2009. "Sexual Deceit Helps Lady-Boy Lizards Mate." *News in Science*, March 3. http://www.abc.net.au/science/articles/2009/03/03/2506127.htm.

Index

A Note on the Author

Dale Peterson has been writing books on science and natural history since completing a Ph.D. in English at Stanford University in 1977. He has traveled through South America, Africa, and into South and Southeast Asia to learn about the lives of wild animals. His recent books include *Elephant Reflections*; *Jane Goodall*; *Eating Apes*; *Storyville, USA*; *Demonic Males* (coauthored with Richard Wrangham); *Chimpanzee Travels*; *Visions of Caliban* (coauthored with Jane Goodall); and *The Deluge and the Ark*. His books have been recognized as Best of the Year by the *Boston Globe*, *Denver Post*, *Discover*, the *Economist*, *Globe and Mail*, *Library Journal*, and the *Village Voice*—and twice distinguished as Notable Book of the Year by the *New York Times*.